Calculus I

Mehdi Rahmani-Andebili

Calculus I

Practice Problems, Methods, and Solutions

Second Edition

 Springer

Mehdi Rahmani-Andebili
Electrical Engineering Department
Arkansas Tech University
Russellville, AR, USA

ISBN 978-3-031-45027-3   ISBN 978-3-031-45028-0   (eBook)
https://doi.org/10.1007/978-3-031-45028-0

This Springer imprint is published by the registered company Springer Nature Switzerland AG
The registered company address is: Gewerbestrasse 11, 6330 Cham, Switzerland

Paper in this product is recyclable.

Calculus is one of the most important courses of many majors, including engineering and science, and even some non-engineering majors like economics and business, which is taught in three successive courses at universities and colleges worldwide. Moreover, in many universities and colleges, a precalculus course is mandatory for under-prepared students as the prerequisite course of Calculus 1.

Unfortunately, some students do not have a solid background and knowledge in math and calculus when they start their education in universities or colleges. This issue prevents them from learning calculus-based courses such as physics and engineering courses. Sometimes, the problem escalates, so they give up and leave the university. Based on my real professorship experience, students do not have a serious issue comprehending physics and engineering courses. In fact, it is the lack of enough knowledge of calculus that hinders them from understanding those courses.

Therefore, a series of calculus textbooks covering Precalculus, Calculus 1, Calculus 2, and Calculus 3 have been prepared to help students succeed in their major. This book, *Calculus 1: Practice Problems, Methods, and Solutions*, is the second edition of the book *Calculus: Practice Problems, Methods, and Solutions*, which was published in 2021. In the new version of the book, many new problems have been added to each chapter. The subjects of the calculus series books are as follows.

Precalculus: Practice Problems, Methods, and Solution

- *Real Number Systems, Exponents and Radicals, and Absolute Values and Inequalities*
- *Systems of Equations*
- *Quadratic Equations*
- *Functions, Algebra of Functions, and Inverse Functions*
- *Factorization of Polynomials*
- *Trigonometric and Inverse Trigonometric Functions*
- *Arithmetic and Geometric Sequences*

Calculus 1: Practice Problems, Methods, and Solution

- *Characteristics of Functions*
- *Trigonometric Equations and Identities*
- *Limits and Continuities*
- *Derivatives and Their Applications*
- *Definite and Indefinite Integrals*

Calculus 2: Practice Problems, Methods, and Solution

- *Applications of Integration*
- *Sequences and Series and Their Applications*
- *Polar Coordinate System*
- *Complex Numbers*

Calculus 3: Practice Problems, Methods, and Solution

- *Linear Algebra and Analytical Geometry*
- *Lines, Surfaces, and Vector Functions in Three-Dimensional Coordinate System*
- *Multivariable Functions*
- *Double Integrals and their Applications*
- *Triple Integrals and their Applications*
- *Line Integrals and Their Applications*

The textbooks include basic and advanced calculus problems with very detailed problem solutions. They can be used as practicing study guides by students and as supplementary teaching sources by instructors. Since the problems have very detailed solutions, the textbooks are helpful for under-prepared students. In addition, they are beneficial for knowledgeable students because they include advanced problems.

In preparing the problems and solutions, care has been taken to use methods typically found in the primary instructor-recommended textbooks. By considering this key point, the textbooks are in the direction of instructors' lectures, and the instructors will not see any untaught and unusual problem solutions in their students' answer sheets.

To help students study in the most efficient way, the problems have been categorized into nine different levels. In this regard, for each problem, a difficulty level (easy, normal, or hard) and a calculation amount (small, normal, or large) have been assigned. Moreover, problems have been ordered in each chapter from the easiest problem with the smallest calculations to the most difficult problems with the largest ones. Therefore, students are suggested to start studying the textbooks from the easiest problems and continue practicing until they reach the normal and then the hardest ones. This classification can also help instructors choose their desirable problems to conduct a quiz or a test. Moreover, the classification of computation amount can help students manage their time during future exams, and instructors assign appropriate problems based on the exam duration.

Russellville, AR, USA   Mehdi Rahmani-Andebili

The author has already published the books and textbooks below with Springer Nature.

Precalculus (2nd Ed.) – Practice Problems, Methods, and Solutions, *Springer Nature*, 2023.

Calculus III – Practice Problems, Methods, and Solutions, *Springer Nature*, 2023.

Calculus II – Practice Problems, Methods, and Solutions, *Springer Nature*, 2023.

Calculus I (2nd Ed.) – Practice Problems, Methods, and Solutions, *Springer Nature*, 2023.

Planning and Operation of Electric Vehicles in Smart Grid, *Springer Nature*, 2023.

Applications of Artificial Intelligence in Planning and Operation of Smart Grid, *Springer Nature*, 2022.

AC Electric Machines – Practice Problems, Methods, and Solutions, *Springer Nature*, 2022.

DC Electric Machines, Electromechanical Energy Conversion Principles, and Magnetic Circuit Analysis- Practice Problems, Methods, and Solutions, *Springer Nature*, 2022.

Applications of Fuzzy Logic in Planning and Operation of Smart Grids, *Springer Nature*, 2021.

Differential Equations – Practice Problems, Methods, and Solutions, *Springer Nature*, 2022.

Feedback Control Systems Analysis and Design – Practice Problems, Methods, and Solutions, *Springer Nature*, 2022.

Power System Analysis – Practice Problems, Methods, and Solutions, *Springer Nature*, 2022.

Advanced Electrical Circuit Analysis – Practice Problems, Methods, and Solutions, *Springer Nature*, 2022.

Design, Control, and Operation of Microgrids in Smart Grids, *Springer Nature*, 2021.

Applications of Fuzzy Logic in Planning and Operation of Smart Grids, *Springer Nature*, 2021.

Operation of Smart Homes, *Springer Nature*, 2021.

AC Electrical Circuit Analysis – Practice Problems, Methods, and Solutions, *Springer Nature*, 2021.

Calculus – Practice Problems, Methods, and Solutions, *Springer Nature*, 2021.

Precalculus – Practice Problems, Methods, and Solutions, *Springer Nature*, 2021.

DC Electrical Circuit Analysis – Practice Problems, Methods, and Solutions, *Springer Nature*, 2020.

Planning and Operation of Plug-in Electric Vehicles: Technical, Geographical, and Social Aspects, *Springer Nature*, 2019.

Contents

Abstract

In this chapter, the basic and advanced problems of functions and inverse functions, algebra of functions, characteristics of functions such as domain of functions, range of functions, axis of symmetry of functions, and types of functions in terms of being odd or even are presented. To help students study the chapter in the most efficient way, the problems are categorized in different levels based on their difficulty levels (easy, normal, and hard) and calculation amounts (small, normal, and large). Moreover, the problems are ordered from the easiest problem with the smallest computations to the most difficult problems with the largest calculations.

1.1 Determine the reflection of the graph of the function below with respect to the origin [1, 2].

$$y = \log \frac{x-1}{x+1}$$

Difficulty level   ● Easy   ○ Normal   ○ Hard
Calculation amount   ● Small   ○ Normal   ○ Large

1) $y = \log \dfrac{x-1}{x+1}$

2) $y = \log \dfrac{1-x}{1+x}$

3) $y = \log \dfrac{1+x}{1-x}$

4) $y = \log \dfrac{x+1}{x-1}$

1.2 Determine the relation of $f(g(h(x)))$ for the information below.

$$f(x) = \ln x^9$$

$$g(x) = \sqrt[3]{x^2}$$

$$h(x) = e^x$$

Difficulty level   ● Easy   ○ Normal   ○ Hard
Calculation amount   ● Small   ○ Normal   ○ Large

1) $6x$

2) $9x$

3) $10x$

4) e^{9x}

1.3 If $f\left(\frac{1}{x}\right) = \sqrt{\frac{2x-1}{x^2}}$ and $g(x) = 2\cos^2(x)$, calculate the value of $fog\left(\frac{\pi}{3}\right)$.

1) 0
2) $\frac{1}{2}$
3) $\frac{\sqrt{3}}{2}$
4) 2

1.4 Determine the value of $fog(3)$ if $f(x) = \sqrt{x} - 2$ and $g(x) = x + 1$.

1) 0
2) 1
3) 2
4) 3

1.5 If $f(x) = 2x - 2$ and $g(x) = x^2 - 1$, solve the equation of $fog(x) = 0$.

1) $\pm\sqrt{2}$
2) ±2
3) $\pm\sqrt{3}$
4) $\sqrt{2}$

1.6 In the function below, calculate the value of $f(-2) + f(2)$.

$$f(x) = \begin{cases} 2x^2 + 4 & x \geq 2 \\ [x] - 4 & x < 2 \end{cases}$$

1) 8
2) 6
3) 10
4) 5

1.7 Determine the domain of $y = \sqrt{2 - x^2}$.

1) $x \leq -\sqrt{2}, x \geq \sqrt{2}$
2) $-\sqrt{2} \leq x \leq \sqrt{2}$
3) $x = 0$
4) $-\sqrt{2} < x < \sqrt{2}$

1.8 Calculate $fog(x)$ if $f(x) = 1 - x^2$ and $g(x) = \sin(x)$.
Difficulty level   ● Easy   ○ Normal   ○ Hard
Calculation amount   ● Small   ○ Normal   ○ Large
1) $\cos^2(x)$
2) $\cos(x)$
3) $\sin(1 - x^2)$
4) $\sin(\cos(x))$

1.9 What is the inverse function of $f(x) = \dfrac{1}{x}$?
Difficulty level   ● Easy   ○ Normal   ○ Hard
Calculation amount   ● Small   ○ Normal   ○ Large
1) x
2) $\dfrac{1}{x}$
3) $-\dfrac{1}{x}$
4) $\sqrt{x}$

1.10 Which one of the following choices is correct about the graph of the relation below?

$$\sinh(xy) - \cosh(xy) + 1 = 0$$

Difficulty level   ● Easy   ○ Normal   ○ Hard
Calculation amount   ○ Small   ● Normal   ○ Large
1) The graph is symmetric with respect to y-axis.
2) The graph is symmetric with respect to the origin and y-axis.
3) The graph is symmetric with respect to x-axis.
4) The graph is symmetric with respect to the origin.

1.11 Calculate $fof(x)$ if $f(x) = \dfrac{1 - x}{1 + x}$.
Difficulty level   ● Easy   ○ Normal   ○ Hard
Calculation amount   ○ Small   ● Normal   ○ Large
1) $\left(\dfrac{1 + x}{1 - x}\right)^2$
2) 1
3) x
4) $\left(\dfrac{1 - x}{1 + x}\right)^2$

1.12 Calculate the inverse function of the following function if $x \geq 1$.

$$f(x) = x^4 - 2x^2 + 1$$

Difficulty level   ○ Easy   ● Normal   ○ Hard
Calculation amount   ● Small   ○ Normal   ○ Large
1) $f^{-1}(x) = \sqrt{1 + \sqrt{x}}$
2) $f^{-1}(x) = \sqrt{1 - \sqrt{x}}$
3) $f^{-1}(x) = -\sqrt{1 + \sqrt{x}}$
4) $f^{-1}(x) = -\sqrt{1 - \sqrt{x}}$

1.13 Calculate the inverse function of the function below.

$$f(x) = \frac{2e^x + 1}{e^x - 3}$$

1) $f^{-1}(x) = \dfrac{3x - 1}{x + 2}$

2) $f^{-1}(x) = \dfrac{x - 2}{3x - 1}$

3) $f^{-1}(x) = \dfrac{x - 2}{3x + 1}$

4) $f^{-1}(x) = \dfrac{3x + 1}{x - 2}$

1.14 Consider the function below.

$$f(x) = \frac{4x + 5}{2x - 3}$$

Which one of the following choices includes a quantity that is not in the domain of $f^{-1}(x)$?

1) -2

2) 2

3) $\dfrac{5}{3}$

4) $\dfrac{3}{2}$

1.15 Calculate the value of $\cos(\pi \sinh(\ln 3))$.

1) -1

2) $-\dfrac{1}{2}$

3) $\dfrac{\sqrt{2}}{2}$

4) $\dfrac{\sqrt{3}}{2}$

1.16 Determine the inverse function of $f(x) = x^2 - 2x$

1) $1 + \sqrt{x + 1}$

2) $1 - \sqrt{x + 1}$

3) $1 + \sqrt{x - 1}$

4) $1 - \sqrt{x - 1}$

1.17 Determine the value of $f(x)$ if $f(x + 1) = x^2 - 2x + 1$.

    1) $(x - 2)^2$
    2) $(x - 1)^2$
    3) $x^2 - 2x$
    4) $(x + 2)^2$

1.18 Which one of the terms below is not a function?
    Difficulty level   ○ Easy   ● Normal ○ Hard
    Calculation amount ● Small ○ Normal ○ Large
    1) $y^2 = x$
    2) $y^3 = x$
    3) $y = \sqrt{x^2}$
    4) $y = \begin{cases} x^2 & x \geq 0 \\ 1 & x < 0 \end{cases}$

1.19 If $f(\sqrt{x}) = x + \sqrt{x}$, calculate the value of $f(2) + f(1)$.
    Difficulty level   ○ Easy   ● Normal ○ Hard
    Calculation amount ● Small ○ Normal ○ Large
    1) 6
    2) 7
    3) 8
    4) 9

1.20 Calculate the value of $f(f(0))$ if:

$$f(x) = \begin{cases} x^2 + 1 & x \geq 1 \\ 2x + 3 & x < 1 \end{cases}$$

    Difficulty level   ○ Easy   ● Normal ○ Hard
    Calculation amount ● Small ○ Normal ○ Large
    1) 3
    2) 5
    3) 10
    4) 26

1.21 Determine the domain of the function below.

$$f(x) = \sqrt{\frac{1 - |x|}{1 + |x|}}$$

    Difficulty level   ○ Easy   ● Normal ○ Hard
    Calculation amount ● Small ○ Normal ○ Large
    1) $\mathbb{R}$
    2) $x \leq 1$
    3) $x \geq 1$
    4) $-1 \leq x \leq 1$

1.22 Determine the domain of the function below.

$$f(x) = \sqrt{\frac{x-1}{x-3}} + \sqrt{\frac{2-x}{x}}$$

Difficulty level   ○ Easy   ● Normal   ○ Hard
Calculation amount   ● Small   ○ Normal   ○ Large
1) $(0,1]$
2) $[0,1]$
3) $(0,2]$
4) $(1,3)$

1.23 Determine the domain of the function below.

$$f(x) = \frac{\sqrt{x}}{|x|-1}$$

Difficulty level   ○ Easy   ● Normal   ○ Hard
Calculation amount   ● Small   ○ Normal   ○ Large
1) $\mathbb{R}$
2) $[0,\infty) - \{1\}$
3) $\mathbb{R} - \{1\}$
4) $[0,\infty)$

1.24 Determine the domain of the function below.

$$f(x) = \frac{\sqrt{x}+1}{x\sqrt{x}+1}$$

Difficulty level   ○ Easy   ● Normal   ○ Hard
Calculation amount   ● Small   ○ Normal   ○ Large
1) $\mathbb{R} - \{0\}$
2) $[1,\infty)$
3) $[0,\infty)$
4) $\mathbb{R} - \{-1,0\}$

1.25 Which number does not exist in the domain of the function below?

$$f(x) = \frac{1-x}{4x+x^3}$$

Difficulty level   ○ Easy   ● Normal   ○ Hard
Calculation amount   ● Small   ○ Normal   ○ Large
1) -2
2) 1
3) 2
4) 0

1.26 Determine $g(x)$ if $f(x) = 2x$ and $f(g(x)) = 2x + 2$.
Difficulty level   ○ Easy   ● Normal   ○ Hard
Calculation amount   ● Small   ○ Normal   ○ Large

    1) $x - 1$
    2) $x + 2$
    3) $x + 1$
    4) $x - 2$

1.27 Determine $g(x)$ if $f(x) = x - 1$ and $f(g(x)) = x$.
    Difficulty level   ○ Easy   ● Normal ○ Hard
    Calculation amount ● Small ○ Normal ○ Large
    1) $x + 1$
    2) $x^2 - x$
    3) $2x - 1$
    4) $x^2 - 1$

1.28 Calculate the inverse function of $f(x) = \sqrt{1 - x}$.
    Difficulty level   ○ Easy   ● Normal ○ Hard
    Calculation amount ● Small ○ Normal ○ Large
    1) $f^{-1}(x) = 1 - x^2,\ x \geq 0$
    2) $f^{-1}(x) = \dfrac{1}{\sqrt{1-x}}$
    3) $f^{-1}(x) = \sqrt{1+x}$
    4) $f^{-1}(x) = 1 - x^2$

1.29 What is the inverse function of $f(x) = \sin(x) - 2$.
    Difficulty level   ○ Easy   ● Normal ○ Hard
    Calculation amount ● Small ○ Normal ○ Large
    1) $2\mathrm{arc}(\sin(x))$
    2) $-2\mathrm{arc}(\sin(x))$
    3) $\mathrm{arc}(\sin(x - 2))$
    4) $\mathrm{arc}(\sin(x + 2))$

1.30 Calculate the inverse function of $fog(x)$ if $f(x) = 3x - 2$ and $g(x) = 2 + x$.
    Difficulty level   ○ Easy   ● Normal ○ Hard
    Calculation amount ● Small ○ Normal ○ Large
    1) $\dfrac{1}{3}x - \dfrac{4}{3}$
    2) $3x - 4$
    3) $\dfrac{1}{3}x + \dfrac{4}{3}$
    4) $3x + 4$

1.31 Calculate the inverse function of $f(x) = x^3 + 3x^2 + 3x + 2$.
    Difficulty level   ○ Easy   ● Normal ○ Hard
    Calculation amount ● Small ○ Normal ○ Large
    1) $1 - \sqrt[3]{x - 1}$
    2) $1 - \sqrt[3]{x + 1}$
    3) $-1 + \sqrt[3]{x - 1}$
    4) $-1 - \sqrt[3]{x + 1}$

1.32 Calculate the value of $\tanh(\ln x)$ in which $x > 0$.
    Difficulty level   ○ Easy   ● Normal ○ Hard
    Calculation amount ○ Small ● Normal ○ Large

1) -1

2) 1

3) $\dfrac{x^2 - 1}{x^2 + 1}$

4) $\dfrac{x^2}{x^2 + 1}$

1.33 If $f(x) + xf(-x) = x^2 + 1$, then what is the value of $f(2)$?

Difficulty level   ○ Easy   ● Normal   ○ Hard

Calculation amount   ○ Small   ● Normal   ○ Large

1) -1

2) -2

3) 3

4) 4

1.34 Calculate the value of $f(f(f(2\cos(x))))$ if $f(x) = x^2 - 2$.

Difficulty level   ○ Easy   ● Normal   ○ Hard

Calculation amount   ○ Small   ● Normal   ○ Large

1) $2\sin^8(x)$

2) $2\cos^8(x)$

3) $2\sin(8x)$

4) $2\cos(8x)$

1.35 Determine the domain of the following function.

$$f\left(\frac{x-1}{x}\right) = \sqrt{2x - 1}$$

Difficulty level   ○ Easy   ● Normal   ○ Hard

Calculation amount   ○ Small   ● Normal   ○ Large

1) $[-1, 0)$

2) $[-1, 1]$

3) $[-1, 1)$

4) $[1, \infty)$

1.36 Determine the domain of $f(x) = \sqrt{1 - \sqrt{x - 1}}$.

Difficulty level   ○ Easy   ● Normal   ○ Hard

Calculation amount   ○ Small   ● Normal   ○ Large

1) $x \geq 1$

2) $1 \leq x \leq 2$

3) $x \leq 2$

4) $\dfrac{5}{4} \leq x \leq \dfrac{7}{4}$

1.37 Determine the domain of the function of $f(x) = \sqrt{|x| - 1} + \sqrt{|x| + 1}$

Difficulty level   ○ Easy   ● Normal   ○ Hard

Calculation amount   ○ Small   ● Normal   ○ Large

1) $\mathbb{R} - [-1, 1]$

2) $\mathbb{R}$

3) $[-1, 1]$

4) $\mathbb{R} - (-1, 1)$

1.38 Calculate the value of $f(3)$ if:

$$f\left(x + \frac{1}{x}\right) = x^2 + \frac{1}{x^2}$$

Difficulty level   ○ Easy   ● Normal   ○ Hard
Calculation amount   ○ Small   ● Normal   ○ Large

1) $\dfrac{28}{3}$

2) $\dfrac{1}{7}$

3) 7

4) $\dfrac{3}{28}$

1.39 For what value of a, the function of $f(x) = |x + 2| + a|x - 2|$ is even?
Difficulty level   ○ Easy   ● Normal   ○ Hard
Calculation amount   ○ Small   ● Normal   ○ Large

1) −1
2) 0
3) 1
4) 2

1.40 The function of $f(x) = x^2 + (A - 1)x$ and $g(x) = (B + 2)x^2 + \sin(x)$ are even and odd functions, respectively. Calculate the value of $A + B$.
Difficulty level   ○ Easy   ● Normal   ○ Hard
Calculation amount   ○ Small   ○ Normal   ● Large

1) −2
2) −1
3) 1
4) 2

1.41 Which one of the following functions is odd?
Difficulty level   ○ Easy   ● Normal   ○ Hard
Calculation amount   ○ Small   ○ Normal   ● Large

1) $\text{arc}(\cos(x))$
2) $\sqrt{1 - x} - \sqrt{1 + x}$
3) $x^4 + x$
4) $x \sin(x)$

1.42 Which one of the functions below is odd?
Difficulty level   ○ Easy   ● Normal   ○ Hard
Calculation amount   ○ Small   ○ Normal   ● Large

1) $|x - 1| + |x + 1|$
2) $\sin(|x|)$
3) $x^3 + x^2$
4) $|x - 1| - |x + 1|$

1.43 Which one of the functions below is even?
Difficulty level   ○ Easy   ● Normal   ○ Hard
Calculation amount   ○ Small   ○ Normal   ● Large

1) $|x - 1| + |x + 1| + |x|$
2) $(x + 1)^4$
3) $f^2(x) + \sqrt[3]{x - 1} = 0$
4) $f(x) = [x] + 1$

1.44 We know that $f(g(x)) = x^2 + \dfrac{1}{x^2} - 4$ and $g(x) = x - \dfrac{1}{x}$. Determine $f(x)$.

Difficulty level   ○ Easy   ○ Normal   ● Hard
Calculation amount   ● Small   ○ Normal   ○ Large
1) $x^2 - 4$
2) $x^2 - 2$
3) x^2
4) $x^2 + 2$

1.45 Calculate the value of $f(-f(x))$ if $f(x) = \begin{cases} x^2 + 1 & x > 0 \\ 1 & x \le 0 \end{cases}$.

Difficulty level   ○ Easy   ○ Normal   ● Hard
Calculation amount   ● Small   ○ Normal   ○ Large
1) 1
2) $x + 1$
3) $x^2 + 1$
4) $(x^2 + 1)^2 + 1$

1.46 For what value of the parameter of "a," the graph of the term below is symmetric with respect to the line of $y = x$.

$$3x^2 + 4xy + (2a - 1)y^2 + (a^2 - 4)x = 7$$

Difficulty level   ○ Easy   ○ Normal   ● Hard
Calculation amount   ○ Small   ● Normal   ○ Large
1) -1
2) -2
3) $-2, 2$
4) 2

1.47 Calculate the value of $gog(x)$ for $x = \sqrt{2} - 1$ based on the following information:

$$f(x) = \cos x$$

$$gof(x) = 1 + \tan^2 x$$

Difficulty level   ○ Easy   ○ Normal   ● Hard
Calculation amount   ○ Small   ● Normal   ○ Large
1) $3 - 2\sqrt{2}$
2) $5\sqrt{2} - 7$
3) $13 - 9\sqrt{2}$
4) $17 - 12\sqrt{2}$

1.48 Determine the domain of the function below.

$$f(x) = \sqrt{\log\left(\frac{5x - x^2}{4}\right)}$$

Difficulty level   ○ Easy   ○ Normal   ● Hard
Calculation amount   ○ Small   ● Normal   ○ Large
1) $1 < x < 4$
2) $0 < x < 5$

3) $1 \leq x \leq 4$

4) $0 \leq x \leq 5$

1.49 Determine the relation of $g\left(\frac{1}{x}\right)$ for the information below.

$$f(x) = \frac{2x}{x+2}$$

$$g(f(x)) = x$$

Difficulty level   ○ Easy   ○ Normal   ● Hard

Calculation amount   ○ Small   ● Normal   ○ Large

1) $\dfrac{2x}{2-x}$

2) $\dfrac{2}{2x-1}$

3) $\dfrac{x-2}{2x}$

4) $\dfrac{2}{1+2x}$

1.50 Determine the relation of $f(x)$ if we have:

$$f\left(\frac{1-\cos 2x}{1+\cos 2x}\right) = \cot x$$

Difficulty level   ○ Easy   ○ Normal   ● Hard

Calculation amount   ○ Small   ● Normal   ○ Large

1) $\sqrt{x}$

2) $\dfrac{1}{\sqrt{x}}$

3) x

4) $\dfrac{1}{x}$

1.51 Determine the domain of $f(x) = \sqrt{\log_x(x^2 + 9)}$.

Difficulty level   ○ Easy   ○ Normal   ● Hard

Calculation amount   ○ Small   ● Normal   ○ Large

1) $(-\infty, \infty)$

2) $(0, \infty)$

3) $[-3, 3]$

4) $(0, \infty) - \{1\}$

1.52 Calculate the range of the function of $f(x) = 2x - 2[x] + 1$.

Difficulty level   ○ Easy   ○ Normal   ● Hard

Calculation amount   ○ Small   ● Normal   ○ Large

1) $[0, 2]$

2) $[1, 3)$

3) $[0, 2)$

4) $[0, 3]$

1.53 Calculate the range of $fog\ (x)$ if $f(x) = x^2 + 1$ and $g(x) = \sqrt{x-1}$.

Difficulty level   ○ Easy   ○ Normal   ● Hard
Calculation amount   ○ Small   ● Normal   ○ Large
1) $[0, \infty)$
2) $[1, \infty)$
3) $[-1, \infty)$
4) $\mathbb{R}$

1.54 Calculate the range of $f(x) = \sqrt{x^2 - 2x + 3}$.

Difficulty level   ○ Easy   ○ Normal   ● Hard
Calculation amount   ○ Small   ● Normal   ○ Large
1) $\left[\sqrt{2}, \infty\right)$
2) $\left[\sqrt{3}, \infty\right)$
3) $[0, \infty)$
4) $[1, \infty)$

1.55 Which one of the following functions is equivalent to $f(x) = |x - 2|$?

Difficulty level   ○ Easy   ○ Normal   ● Hard
Calculation amount   ○ Small   ○ Normal   ● Large
1) $g_1(x) = \left| \dfrac{x^2 - 3x + 2}{x - 1} \right|$

2) $g_2(x) = \left| \dfrac{x^2 - 4}{x + 2} \right|$

3) $g_3(x) = \dfrac{(x-2)^2}{|x-2|}$

4) $g_4(x) = \dfrac{|6x - 12|}{6}$

1.56 Which one of the choices is the axis of symmetry of the following function?

$$f(x) = \sqrt[3]{x} - \sqrt[3]{x+2}$$

Difficulty level   ○ Easy   ○ Normal   ● Hard
Calculation amount   ○ Small   ○ Normal   ● Large
1) $x = -2$
2) $x = -1$
3) $x = 1$
4) $x = 2$.

References

1. Rahmani-Andebili, M. (2021). Calculus – Practice Problems, Methods, and Solutions, Springer Nature, 2021.
2. Rahmani-Andebili, M. (2021). Precalculus – Practice Problems, Methods, and Solutions, Springer Nature, 2021.

Abstract

In this chapter, the problems of the first chapter are fully solved, in detail, step-by-step, and with different methods.

2.1 Based on the information given in the problem, the function is as follows [1, 2]:

$$y = \log \frac{x-1}{x+1}$$

The reflection of the graph of a function in the form of $f(x, y)$ with respect to the origin can be achieved by changing x to $-x$ and y to $-y$. In other words, $f(-x, -y)$ is the reflection of $f(x, y)$ with respect to the origin. Therefore:

$$-y = \log \frac{-x-1}{-x+1}$$

$$\Rightarrow -y = \log \frac{x+1}{x-1} \Rightarrow y = -\log \frac{x+1}{x-1} \Rightarrow y = \log \frac{x-1}{x+1}$$

In the calculations, the rule below was used.

$$-\log a = \log \frac{1}{a}$$

Choice (1) is the answer.

2.2 Based on the information given in the problem, we have:

$$f(x) = \ln x^9$$

$$g(x) = \sqrt[3]{x^2}$$

$$h(x) = e^x$$

The problem can be solved as follows:

$$f(g(h(x))) = f(g(e^x)) = f\left(\sqrt[3]{(e^x)^2}\right) = f\left(e^{\frac{2}{3}x}\right) = \ln \left(e^{\frac{2}{3}x}\right)^9 = \ln e^{6x}$$

© The Author(s), under exclusive license to Springer Nature Switzerland AG 2023
M. Rahmani-Andebili, *Calculus I*, https://doi.org/10.1007/978-3-031-45028-0_2

$$\Rightarrow f(g(h(x))) = 6x$$

In the calculations, the rules below were used:

$$\sqrt[m]{(f(x))^n} = (f(x))^{\frac{n}{m}}$$

$$\ln e^{f(x)} = f(x)$$

Choice (1) is the answer.

2.3 Based on the information given in the problem, we have:

$$f\left(\frac{1}{x}\right) = \sqrt{\frac{2x-1}{x^2}}$$

$$g(x) = 2\cos^2(x)$$

The problem can be solved as follows:

$$fog\left(\frac{\pi}{3}\right) = f\left(g\left(\frac{\pi}{3}\right)\right) = f\left(2\cos^2\left(\frac{\pi}{3}\right)\right) = f\left(2\left(\frac{1}{2}\right)^2\right) = f\left(\frac{1}{2}\right) = \sqrt{\frac{2(2)-1}{2^2}}$$

$$\Rightarrow fog\left(\frac{\pi}{3}\right) = \frac{\sqrt{3}}{2}$$

Choice (3) is the answer.

2.4 Based on the information given in the problem, we have:

$$f(x) = \sqrt{x} - 2$$

$$g(x) = x + 1$$

The problem can be solved as follows:

$$fog(3) = f(g(3)) = f(3+1) = f(4) = \sqrt{4} - 2$$

$$\Rightarrow fog(3) = 0$$

Choice (1) is the answer.

2.5 Based on the information given in the problem, we have:

$$f(x) = 2x - 2$$

$$g(x) = x^2 - 1$$

The problem can be solved as follows:

$$fog(x) = f(g(x)) = f(x^2 - 1) = 2(x^2 - 1) - 2 = 2x^2 - 4$$

$$fog(x) = 0 \Rightarrow 2x^2 - 4 = 0 \Rightarrow x^2 = 2$$

$$\Rightarrow x = \pm\sqrt{2}$$

Choice (1) is the answer.

2.6 Based on the information given in the problem, we have:

$$f(x) = \begin{cases} 2x^2 + 4 & x \geq 2 \\ [x] - 4 & x < 2 \end{cases}$$

The problem can be solved as follows:

$$f(2) = 2(2)^2 + 4 = 12$$

$$f(-2) = [-2] - 4 = -6$$

$$\Rightarrow f(2) + f(-2) = 12 + (-6)$$

$$\Rightarrow f(2) + f(-2) = 6$$

Choice (2) is the answer.

2.7 Based on the information given in the problem, we have:

$$y = \sqrt{2 - x^2}$$

The domain of a function in the radical form with an even root is determined by considering the radicand equal and greater than zero. Therefore:

$$2 - x^2 \geq 0 \Rightarrow x^2 \leq 2$$

$$\Rightarrow -\sqrt{2} \leq x \leq \sqrt{2}$$

Choice (2) is the answer.

2.8 From trigonometry, we know that:

$$\sin^2(x) + \cos^2(x) = 1$$

Based on the information given in the problem, we have:

$$f(x) = 1 - x^2$$

$$g(x) = \sin(x)$$

Therefore:

$$fog(x) = f(g(x)) = f(\sin(x)) = 1 - (\sin(x))^2$$

$$\Rightarrow fog(x) = \cos^2(x)$$

Choice (1) is the answer.

2.9 Based on the information given in the problem, we have:

$$f(x) = \frac{1}{x}$$

To calculate the inverse function of a function, we need to determine x based on y. After that, we must replace x by y and vice versa. Note that the domain of $f^{-1}(x)$ is the same as the range of $f(x)$.

Therefore:

$$y = \frac{1}{x} \Rightarrow x = \frac{1}{y} \Rightarrow y = \frac{1}{x}$$

$$\Rightarrow f^{-1}(x) = \frac{1}{x}$$

Choice (2) is the answer.

2.10 Based on the information given in the problem, we have:

$$\sinh(xy) - \cosh(xy) + 1 = 0$$

A function in the form of $f(x, y) = 0$ is symmetric with respect to x-axis if it does not change by the conversion of $y \rightarrow -y$, that is, $f(x, y) = f(x, -y)$.

Moreover, a function in the form of $f(x, y) = 0$ is symmetric with respect to y-axis if it does not change by the conversion of $x \rightarrow -x$, that is, $f(x, y) = f(-x, y)$.

In addition, a function in the form of $f(x, y) = 0$ is symmetric with respect to the origin if it does not change by the conversions of $x \rightarrow -x$ and $y \rightarrow -y$, that is, $f(x, y) = f(-x, -y)$.

Therefore:

$$y \rightarrow -y \Rightarrow \sinh(-xy) - \cosh(-xy) + 1 = 0$$

$$\Rightarrow -\sinh(xy) - \cosh(xy) + 1 = 0$$

$$\Rightarrow f(x, y) \neq f(x, -y)$$

Therefore, the relation is not symmetric with respect to x-axis.

$$x \rightarrow -x \Rightarrow \sinh(-xy) - \cosh(-xy) + 1 = 0$$

$$\Rightarrow -\sinh(xy) - \cosh(xy) + 1 = 0$$

$$\Rightarrow f(x, y) \neq f(-x, y)$$

Thus, the relation is not symmetric with respect to y-axis.

$$\begin{cases} x \longrightarrow -x \\ y \longrightarrow -y \end{cases} \Rightarrow \sinh((-x)(-y)) - \cosh((-x)(-y)) + 1 = 0$$

$$\Rightarrow \sinh(xy) - \cosh(xy) + 1 = 0$$

$$\Rightarrow f(x, y) = f(-x, -y)$$

Hence, the relation is symmetric with respect to the origin.

In the calculations, the rules below were used:

$$\sinh(-a) = -\sinh(a)$$

$$\cosh(-a) = \cosh(a)$$

Choice (4) is the answer.

2.11 Based on the information given in the problem, we have:

$$f(x) = \frac{1-x}{1+x}$$

Therefore:

$$fof(x) = f(f(x)) = f\left(\frac{1-x}{1+x}\right) = \frac{1 - \frac{1-x}{1+x}}{1 + \frac{1-x}{1+x}} = \frac{\frac{1+x-(1-x)}{1+x}}{\frac{1+x+1-x}{1+x}} = \frac{\frac{2x}{1+x}}{\frac{2}{1+x}}$$

$$\Rightarrow fof(x) = x$$

Choice (3) is the answer.

2.12 Based on the information given in the problem, we have:

$$y = x^4 - 2x^2 + 1$$

$$x \geq 1$$

To calculate the inverse function of a function, we need to determine x based on y. After that, we must replace x by y and vice versa. Note that the domain of $f^{-1}(x)$ is the same as the range of $f(x)$.

Therefore:

$$y = x^4 - 2x^2 + 1 \Rightarrow y = \left(x^2 - 1\right)^2$$

Since $x \geq 1$, the value of y is positive in the last equation. Hence, its square root can be determined.

$$\sqrt{y} = \left(x^2 - 1\right)$$

$$\Rightarrow 1 + \sqrt{y} = x^2$$

$$\Rightarrow x = \sqrt{1 + \sqrt{y}}$$

$$\Rightarrow y = \sqrt{1 + \sqrt{x}}$$

$$\Rightarrow f^{-1}(x) = \sqrt{1 + \sqrt{x}}$$

Choice (1) is the answer.

2.13 Based on the information given in the problem, we have:

$$y = \frac{2e^x + 1}{e^x - 3}$$

To calculate the inverse function of a function, we need to determine x based on y. After that, we must replace x by y and vice versa. Note that the domain of $f^{-1}(x)$ is the same as the range of $f(x)$.

Therefore:

$$y = \frac{2e^x + 1}{e^x - 3} \Rightarrow ye^x - 3y = 2e^x + 1$$

$$\Rightarrow e^x(y - 2) = 1 + 3y \Rightarrow e^x = \frac{1 + 3y}{y - 2}$$

$$\xrightarrow{\ln} x = \ln \frac{1 + 3y}{y - 2}$$

$$\Rightarrow y = \ln \frac{1 + 3x}{x - 2}$$

$$\Rightarrow f^{-1}(x) = \ln \frac{3x + 1}{x - 2}$$

In the calculations, the rule below was used:

$$\ln e^{f(x)} = f(x)$$

Choice (4) is the answer.

2.14 Based on the information given in the problem, we have:

$$f(x) = \frac{4x + 5}{2x - 3}$$

To calculate the inverse function of a function, we need to determine x based on y. After that, we must replace x by y and vice versa. Note that the domain of $f^{-1}(x)$ is the same as the range of $f(x)$.

Therefore:

$$y = \frac{4x + 5}{2x - 3} \Rightarrow 2xy - 3y = 4x + 5 \Rightarrow x(2y - 4) = 3y + 5$$

$$\Rightarrow x = \frac{3y + 5}{2y - 4}$$

$$\Rightarrow y = \frac{3x + 5}{2x - 4}$$

$$\Rightarrow f^{-1}(x) = \frac{3x + 5}{2x - 4}$$

The domain of $f^{-1}(x)$ can be determined as follows:

$$2x - 4 = 0 \Rightarrow x = 2$$

$$\Rightarrow D_{f^{-1}(x)} = \mathbb{R} - \{2\}$$

Therefore, $x = 2$ is not in the in the domain of $f^{-1}(x)$. Choice (2) is the answer.

2.15 From trigonometry, we know that:

$$\sinh x = \frac{e^x - e^{-x}}{2}$$

Therefore:

$$\sinh(\ln 3) = \frac{e^{\ln 3} - e^{-\ln 3}}{2} = \frac{3 - \frac{1}{3}}{2} = \frac{8}{6} = \frac{4}{3}$$

$$\Rightarrow \cos(\pi \sinh(\ln 3)) = \cos\left(\frac{4}{3}\pi\right) = \cos\left(\pi + \frac{\pi}{3}\right) = -\cos\left(\frac{\pi}{3}\right)$$

$$\Rightarrow \cos(\pi \sinh(\ln 3)) = -\frac{1}{2}$$

In the calculations, the rules below were used:

$$e^{\ln f(x)} = f(x)$$

$$\cos\left(\frac{\pi}{3}\right) = \frac{1}{2}$$

Choice (2) is the answer.

2.16 Based on the information given in the problem, we have:

$$f(x) = x^2 - 2x$$

To calculate the inverse function of a function, we need to determine x based on y. After that, we must replace x by y and vice versa. Note that the domain of $f^{-1}(x)$ is the same as the range of $f(x)$.

First, we need to define the function in a square form as follows:

$$y = x^2 - 2x + 1 - 1 \Rightarrow y = (x - 1)^2 - 1$$

$$\Rightarrow y + 1 = (x - 1)^2 \Rightarrow \sqrt{y + 1} = x - 1 \Rightarrow x = 1 + \sqrt{y + 1}$$

$$\Rightarrow y = 1 + \sqrt{x + 1}$$

$$\Rightarrow f^{-1}(x) = 1 + \sqrt{x + 1}$$

Choice (1) is the answer.

2.17 Based on the information given in the problem, we have:

$$f(x + 1) = x^2 - 2x + 1$$

The problem can be solved as follows:

$$f(x + 1) = (x - 1)^2$$

$$\xrightarrow{x \to x - 1} f((x - 1) + 1) = ((x - 1) - 1)^2$$

$$\Rightarrow f(x) = (x - 2)^2$$

Choice (1) is the answer.

2.18 A mathematical relation is a function if for any value of x, one value of y is achieved at most. Or, a function is a binary relation between two sets that associates every element of the first set to exactly one element of the second set. Herein, $y^2 = x$ is not a function because for $x = 1$, $y = -1, 1$ are achieved.

Choice (1) is the answer.

2.19 Based on the information given in the problem, we have:

$$f\left(\sqrt{x}\right) = x + \sqrt{x}$$

The problem can be solved as follows:

$$f\left(\sqrt{x}\right) = x + \sqrt{x} = \left(\sqrt{x}\right)^2 + \sqrt{x}$$

$$\xrightarrow{\sqrt{x} \to x} f(x) = x^2 + x$$

Now, the function is in a standard form.

$$f(2) + f(1) = 2^2 + 2 + 1^2 + 1$$

$$\Rightarrow f(2) + f(1) = 8$$

Choice (3) is the answer.

2.20 Based on the information given in the problem, we have:

$$f(x) = \begin{cases} x^2 + 1 & x \geq 1 \\ 2x + 3 & x < 1 \end{cases}$$

The problem can be solved as follows:

$$f(0) = 2 \times 0 + 3 = 3$$

$$\Rightarrow f(f(0)) = f(3) = 3^2 + 1$$

$$\Rightarrow f(f(0)) = 10$$

Choice (3) is the answer.

2.21 Based on the information given in the problem, we have:

$$f(x) = \sqrt{\frac{1 - |x|}{1 + |x|}}$$

The domain of a function in radical form, including an even root, is determined by considering the radicand equal and greater than zero. Therefore:

$$\frac{1 - |x|}{1 + |x|} \geq 0$$

$$\xrightarrow{1 + |x| > 0} 1 - |x| \geq 0 \Rightarrow |x| \leq 1 \Rightarrow 1 \leq x \leq 1$$

Choice (4) is the answer.

2.22 Based on the information given in the problem, we have:

$$f(x) = \sqrt{\frac{x-1}{x-3}} + \sqrt{\frac{2-x}{x}}$$

The domain of a function in radical form that includes an even root is determined by considering the radicand equal and greater than zero. Therefore:

$$\begin{cases} \dfrac{x-1}{x-3} \geq 0 \\ \dfrac{2-x}{x} \geq 0 \end{cases} \Rightarrow \begin{cases} \dfrac{x-1}{x-3} \geq 0 \\ \dfrac{x-2}{x} \leq 0 \end{cases} \Rightarrow \begin{cases} x \leq 1, x > 3 \\ 0 < x \leq 2 \end{cases} \xrightarrow{\cap} 0 < x \leq 1$$

Choice (1) is the answer.

2.23 Based on the information given in the problem, we have:

$$f(x) = \frac{\sqrt{x}}{|x| - 1}$$

The domain of a function in radical form, including even root, is determined by considering the radicand equal and greater than zero. Moreover, the value of those x that make the denominator zero must be excluded from the domain. Thus:

$$\begin{cases} x \geq 0 \\ |x| - 1 \neq 0 \end{cases} \Rightarrow \begin{cases} x \geq 0 \\ |x| \neq 1 \end{cases} \Rightarrow \begin{cases} x \geq 0 \\ x \neq \pm 1 \end{cases} \xrightarrow{\cap} D_f = [0, \infty) - \{1\}$$

Choice (2) is the answer.

2.24 Based on the information given in the problem, we have:

$$f(x) = \frac{\sqrt{x}+1}{x\sqrt{x}+1}$$

The domain of a function in radical form, including even root, is determined by considering the radicand equal and greater than zero. Moreover, the value of those x that make the denominator zero must be excluded from the domain. Thus:

$$\begin{cases} x \geq 0 \\ x\sqrt{x}+1 \neq 0 \end{cases}$$

Note that $x\sqrt{x}+1 \neq 0$ is true for any x. Hence:

$$x \geq 0 \Rightarrow D_f = [0, \infty)$$

Choice (3) is the answer.

2.25 Based on the information given in the problem, we have:

$$f(x) = \frac{1-x}{4x+x^3}$$

The value of those x that make the denominator zero are not in the domain. Thus:

$$4x+x^3 \neq 0 \Rightarrow x(4+x^2) \neq 0$$

Note that $4 + x^2 \neq 0$ for any x. Hence:

$$x = 0$$

Choice (4) is the answer.

2.26 Based on the information given in the problem, we have:

$$f(x) = 2x \tag{1}$$

$$f(g(x)) = 2x + 2 \tag{2}$$

Therefore:

$$f(g(x)) = 2g(x) \tag{3}$$

Solving (2) and (3):

$$2g(x) = 2x + 2 \Rightarrow g(x) = x + 1$$

Choice (3) is the answer.

2.27 Based on the information given in the problem, we have:

$$f(x) = x - 1 \tag{1}$$

$$f(g(x)) = x \tag{2}$$

Thus:

$$f(g(x)) = g(x) - 1 \tag{3}$$

Solving (2) and (3):

$$g(x) - 1 = x \Rightarrow g(x) = x + 1$$

Choice (1) is the answer.

2.28 Based on the information given in the problem, we have:

$$f(x) = \sqrt{1 - x}$$

To calculate the inverse function of a function, we need to determine x based on y. After that, we must replace x by y and vice versa. Note that the domain of $f^{-1}(x)$ is the same as the range of $f(x)$.

Therefore:

$$y = \sqrt{1 - x} \Rightarrow y^2 = 1 - x \Rightarrow x = 1 - y^2$$

$$\Rightarrow y = 1 - x^2$$

$$\Rightarrow f^{-1}(x) = 1 - x^2$$

Since the domain of $f^{-1}(x)$ is the same as the range of $f(x)$, which is $[0, \infty)$, we need to add $x \geq 0$ to $f^{-1}(x)$ as its domain. Thus:

$$f^{-1}(x) = 1 - x^2, x \geq 0$$

Choice (1) is the answer.

2.29 Based on the information given in the problem, we have:

$$f(x) = \sin(x) - 2$$

To calculate the inverse function of a function, we need to determine x based on y. After that, we must replace x by y and vice versa. Note that the domain of $f^{-1}(x)$ is the same as the range of $f(x)$.

Therefore:

$$y = \sin(x) - 2 \Rightarrow y + 2 = \sin(x) \Rightarrow x = \arc(\sin(y + 2))$$

$$\Rightarrow y = \arc(\sin(x + 2))$$

$$\Rightarrow f^{-1}(x) = \arc(\sin(x + 2))$$

Choice (4) is the answer.

2.30 Based on the information given in the problem, we have:

$$f(x) = 3x - 2$$

$$g(x) = 2 + x$$

First, we need to determine $fog(x)$ as follows:

$$fog(x) = f(g(x)) = f(2 + x) = 3(2 + x) - 2 = 3x + 4$$

To calculate the inverse function of a function, we need to determine x based on y. After that, we must replace x by y and vice versa. Note that the domain of $f^{-1}(x)$ is the same as the range of $f(x)$.

Therefore:

$$y = 3x + 4 \Rightarrow 3x = y - 4$$

$$\Rightarrow x = \frac{1}{3}y - \frac{4}{3}$$

$$\Rightarrow y = \frac{1}{3}x - \frac{4}{3}$$

$$\Rightarrow f^{-1}(x) = \frac{1}{3}x - \frac{4}{3}$$

Choice (1) is the answer.

2.31 Based on the information given in the problem, we have:

$$f(x) = x^3 + 3x^2 + 3x + 2$$

To calculate the inverse function of a function, we need to determine x based on y. After that, we must replace x by y and vice versa. Note that the domain of $f^{-1}(x)$ is the same as the range of $f(x)$.

First, we need to define the function in a cube form as follows:

$$\Rightarrow f(x) = \left(x^3 + 3x^2 + 3x + 1\right) + 1 = (x + 1)^3 + 1$$

$$\Rightarrow y = (x + 1)^3 + 1 \Rightarrow y - 1 = (x + 1)^3 \Rightarrow (y - 1)^{\frac{1}{3}} = x + 1$$

$$\Rightarrow x = (y - 1)^{\frac{1}{3}} - 1$$

$$\Rightarrow y = (x - 1)^{\frac{1}{3}} - 1$$

$$\Rightarrow f^{-1}(x) = -1 + \sqrt[3]{x - 1}$$

Choice (3) is the answer.

2.32 Based on the information given in the problem, we have:

$$x > 0$$

From trigonometry, we know that:

$$\sinh t = \frac{e^t - e^{-t}}{2}$$

$$\cosh t = \frac{e^t + e^{-t}}{2}$$

$$\tanh t = \frac{\sinh t}{\sinh t}$$

Therefore:

$$\sinh(\ln x) = \frac{e^{\ln x} - e^{-\ln x}}{2} = \frac{x - \frac{1}{x}}{2}$$

$$\cosh(\ln x) = \frac{e^{\ln x} + e^{-\ln x}}{2} = \frac{x + \frac{1}{x}}{2}$$

$$\tanh(\ln x) = \frac{\sinh(\ln x)}{\sinh(\ln x)} = \frac{\frac{x - \frac{1}{x}}{2}}{\frac{x + \frac{1}{x}}{2}} = \frac{x - \frac{1}{x}}{x + \frac{1}{x}}$$

$$\Rightarrow \tanh(\ln x) = \frac{x^2 - 1}{x^2 + 1}$$

In the calculations, the rule below was used:

$$e^{\ln f(x)} = f(x)$$

Choice (3) is the answer.

2.33 Based on the information given in the problem, we have:

$$f(x) + x f(-x) = x^2 + 1$$

The problem can be solved as follows:

$$\xrightarrow{\substack{x = 2 \\ x = -2}} \begin{cases} f(2) + 2f(-2) = 2^2 + 1 \\ f(-2) + (-2)f(2) = (-2)^2 + 1 \end{cases} \Rightarrow \begin{cases} f(2) + 2f(-2) = 5 \\ f(-2) - 2f(2) = 5 \end{cases}$$

$$\xrightarrow[\times(-2)]{\Rightarrow} \begin{cases} f(2) + 2f(-2) = 5 \\ -2f(-2) + 4f(2) = -10 \end{cases}$$

$$\xrightarrow{+} 5f(2) = -5 \Rightarrow f(2) = -1$$

Choice (1) is the answer.

2.34 From trigonometry, we know that:

$$1 + \cos(2x) = 2\cos^2(x)$$

Based on the information given in the problem, we have:

$$f(x) = x^2 - 2$$

The problem can be solved as follows:

$$\Rightarrow f(2\cos(x)) = (2\cos(x))^2 - 2 = 4\cos^2(x) - 2 = 2(1 + \cos(2x)) - 2 = 2\cos(2x)$$

$$\Rightarrow f(f(2\cos(x))) = (2\cos(2x))^2 - 2 = 4\cos^2(2x) - 2 = 2(1 + \cos(4x)) - 2 = 2\cos(4x)$$

$$\Rightarrow f(f(f(2\cos(x)))) = (2\cos(4x))^2 - 2 = 4\cos^2(4x) - 2 = 2(1 + \cos(8x)) - 2$$

$$\Rightarrow f(f(f(2\cos(x)))) = 2\cos(8x)$$

Choice (4) is the answer.

2.35 Based on the information given in the problem, we have:

$$f\left(\frac{x-1}{x}\right) = \sqrt{2x - 1}$$

First, we need to determine the $f(x)$ as follows:

$$\frac{x-1}{x} = t$$

$$\Rightarrow x = \frac{1}{1-t}$$

$$\Rightarrow f(t) = \sqrt{2 \times \frac{1}{1-t} - 1} = \sqrt{\frac{1+t}{1-t}}$$

$$\xrightarrow{t \to x} f(x) = \sqrt{\frac{1+x}{1-x}}$$

The domain of a function in radical form, including even root, is determined by considering the radicand equal and greater than zero. Therefore:

$$\frac{1+x}{1-x} \geq 0 \Rightarrow \frac{x+1}{x-1} \leq 0 \Rightarrow -1 \leq x < 1$$

Note, that $x = 1$ must be excluded from the domain, since it makes the denominator zero. Choice (3) is the answer.

2.36 Based on the information given in the problem, we have:

$$f(x) = \sqrt{1 - \sqrt{x - 1}}$$

The domain of a function in radical form that includes an even root is determined by considering the radicand equal and greater than zero. Therefore:

$$\begin{cases} 1 - \sqrt{x-1} \geq 0 \\ x - 1 \geq 0 \end{cases} \Rightarrow \begin{cases} \sqrt{x-1} \leq 1 \\ x \geq 1 \end{cases} \Rightarrow \begin{cases} x - 1 \leq 1 \\ x \geq 1 \end{cases} \Rightarrow \begin{cases} x \leq 2 \\ x \geq 1 \end{cases} \overset{\cap}{\Rightarrow} 1 \leq x \leq 2$$

Choice (2) is the answer.

2.37 Based on the information given in the problem, we have:

$$f(x) = \sqrt{|x| - 1} + \sqrt{|x| + 1}$$

The domain of a function in radical form, including even root, is determined by considering the radicand equal and greater than zero.

$$\begin{cases} |x| - 1 \geq 0 \\ |x| + 1 \geq 0 \end{cases} \Rightarrow \begin{cases} |x| \geq 1 \\ x \in \mathbb{R} \end{cases} \Rightarrow \begin{cases} x \leq -1, x \geq 1 \\ x \in \mathbb{R} \end{cases} \overset{\cap}{\Rightarrow} x \leq -1, x \geq 1 \Rightarrow D_f = \mathbb{R} - (-1, 1)$$

Note that $|x| + 1 \geq 0$ is true for any x. Choice (4) is the answer.

2.38 Based on the information given in the problem, we have:

$$f\left(x + \frac{1}{x}\right) = x^2 + \frac{1}{x^2}$$

The problem can be solved as follows:

$$\Rightarrow f\left(x + \frac{1}{x}\right) = x^2 + \frac{1}{x^2} + 2 - 2 = \left(x + \frac{1}{x}\right)^2 - 2$$

$$\xrightarrow{\;x + \frac{1}{x} \to x\;} f(x) = x^2 - 2 \Rightarrow f(3) = 3^2 - 2$$

$$\Rightarrow f(3) = 7$$

Choice (3) is the answer.

2.39 Based on the information given in the problem, we have:

$$f(x) = |x + 2| + a|x - 2| \qquad (1)$$

Based on the definition, the function of $f(x)$ is an even function if its domain is symmetric and:

$$f(-x) = f(x) \qquad (2)$$

Therefore:

$$\Rightarrow f(-x) = |-x + 2| + a|-x - 2| = |-(x - 2)| + a|-(x + 2)| = |x - 2| + a|x + 2| \qquad (3)$$

Solving (1), (2), and (3):

$$|x - 2| + a|x + 2| = |x + 2| + a|x - 2| \Rightarrow a = 1$$

Choice (3) is the answer.

2.40 Based on the information given in the problem, we have:

$$f(x) = x^2 + (A-1)x \text{ is an even function} \qquad (1)$$

$$g(x) = (B+2)x^2 + \sin(x) \text{ is an odd function} \qquad (2)$$

Based on the definition, the function of $f(x)$ is an even function if its domain is symmetric and:

$$f(-x) = f(x) \qquad (3)$$

Additionally, the function of $f(x)$ is an odd function if its domain is symmetric and:

$$f(-x) = -f(x) \qquad (4)$$

Solving (1) and (3):

$$(-x)^2 + (A-1)(-x) = x^2 + (A-1)x \Rightarrow 2(A-1)x = 0 \Rightarrow A = 1$$

Solving (2) and (4):

$$(B+2)(-x)^2 + \sin(-x) = -\big((B+2)x^2 + \sin(x)\big)$$

$$\Rightarrow (B+2)x^2 - \sin(x) = -(B+2)x^2 - \sin(x) \Rightarrow 2(B+2)x^2 = 0 \Rightarrow B = -2$$

Therefore:

$$A + B = 1 + (-2) = -1$$

Choice (2) is the answer.

2.41 Based on the definition, the function of $f(x)$ is an odd function if its domain is symmetric and:

$$f(-x) = -f(x)$$

Choice (1):

$$f(x) = \mathrm{arc}(\cos(x))$$

$$\Rightarrow f(-x) = \mathrm{arc}(\cos(-x)) = \pi - \mathrm{arc}(\cos(x)) \neq \{-f(x), f(x)\}$$

$$\Rightarrow \text{Not even nor odd}$$

Choice (2):

$$f(x) = \sqrt{1-x} - \sqrt{1+x}$$

$$\Rightarrow f(-x) = \sqrt{1-(-x)} - \sqrt{1+(-x)} = \sqrt{1+x} - \sqrt{1-x} = -\left(\sqrt{1-x} - \sqrt{1+x}\right) = -f(x)$$

$$\Rightarrow \text{Odd}$$

Choice (3):

$$f(x) = x^4 + x$$

$$\Rightarrow f(-x) = (-x)^4 + (-x) = x^4 - x \neq \{-f(x), f(x)\}$$

$$\Rightarrow \text{Not even nor odd}$$

Choice (4):

$$f(x) = x\sin(x)$$

$$\Rightarrow f(-x) = -x\sin(-x) = x\sin(x) = f(x)$$

$$\Rightarrow \text{Even}$$

Choice (2) is the answer.

2.42 Based on the definition, the function of $f(x)$ is an odd function if its domain is symmetric and:

$$f(-x) = -f(x)$$

Choice (1):

$$f(x) = |x - 1| + |x + 1|$$

$$\Rightarrow f(-x) = |-x - 1| + |-x + 1| = |-(x+1)| + |-(x-1)| = |x+1| + |x-1| = f(x)$$

$$\Rightarrow \text{Even}$$

Choice (2):

$$f(x) = \sin(|x|)$$

$$\Rightarrow f(-x) = \sin(|-x|) = \sin(|x|) = f(x)$$

$$\Rightarrow \text{Even}$$

Choice (3):

$$f(x) = x^3 + x^2 \Rightarrow f(-x) = (-x)^3 + (-x)^2 = -x^3 + x^2 \neq \{-f(x), f(x)\}$$

$$\Rightarrow \text{Not even nor odd}$$

Choice (4):

$$f(x) = |x - 1| - |x + 1|$$

$$\Rightarrow f(-x) = |-x - 1| - |-x + 1| = |-(x+1)| - |-(x-1)| = |x+1| - |x-1| = -f(x)$$

$$\Rightarrow \text{Odd}$$

Choice (4) is the answer.

2.43 Based on the definition, the function of $f(x)$ is an even function if its domain is symmetric and:

$$f(-x) = f(x)$$

Choice (1):

$$f(x) = |x - 1| + |x + 1| + |x|$$

$$\Rightarrow f(-x) = |-x - 1| + |-x + 1| + |-x| = |-(x + 1)| + |-(x - 1)| + |-x| = |x + 1| + |x - 1| + |x| = f(x)$$

$$\Rightarrow \text{Even}$$

Choice (2):

$$f(x) = (x + 1)^4$$

$$\Rightarrow f(-x) = (-x + 1)^4 = (x - 1)^4 \neq \{-f(x), f(x)\}$$

$$\Rightarrow \text{Not even nor odd}$$

Choice (3):

$$f^2(x) + \sqrt[3]{x - 1} = 0 \Rightarrow \text{Not a function}$$

Choice (4):

$$f(x) = [x] + 1$$

$$\Rightarrow f(-x) = [-x] + 1 \neq \{-f(x), f(x)\}$$

$$\Rightarrow \text{Not even nor odd}$$

Choice (1) is the answer.

2.44 Based on the information given in the problem, we have:

$$g(x) = x - \frac{1}{x} \tag{1}$$

$$f(g(x)) = x^2 + \frac{1}{x^2} - 4 \tag{2}$$

The problem can be solved as follows:

$$f(g(x)) = x^2 + \frac{1}{x^2} - 2 - 2 = \left(x - \frac{1}{x}\right)^2 - 2 \tag{3}$$

Solving (1) and (3):

$$f(g(x)) = (g(x))^2 - 2$$

$$\Rightarrow f(x) = x^2 - 2$$

Choice (2) is the answer.

2.45 Based on the information given in the problem, we have:

$$f(x) = \begin{cases} x^2 + 1 & x > 0 \\ 1 & x \le 0 \end{cases}$$

As can be noticed from $f(x)$, the value of function is always positive. Therefore, the value of $-f(x)$ is always negative. Hence:

$$f(-f(x)) = 1$$

Choice (1) is the answer.

2.46 A function in the form of $f(x, y) = 0$ is symmetric with respect to the line of $y = x$ if $f(x, y) = f(y, x)$.

Based on the information given in the problem, we have:

$$3x^2 + 4xy + (2a - 1)y^2 + (a^2 - 4)x = 7$$

$$\Rightarrow f(x, y) = 3x^2 + 4xy + (2a - 1)y^2 + (a^2 - 4)x - 7$$

Moreover, from $f(x, y) = f(y, x)$, we have:

$$\Rightarrow 3x^2 + 4xy + (2a - 1)y^2 + (a^2 - 4)x - 7 = 3y^2 + 4yx + (2a - 1)x^2 + (a^2 - 4)y - 7$$

$$\Rightarrow (4 - 2a)x^2 + (2a - 4)y^2 + (a^2 - 4)x - (a^2 - 4)y = 0$$

$$\Rightarrow \begin{cases} 4 - 2a = 0 \\ 2a - 4 = 0 \\ a^2 - 4 = 0 \\ a^2 - 4 = 0 \end{cases} \Rightarrow a = \{2\} \cap \{2\} \cap \{-2, 2\} \cap \{-2, 2\} \Rightarrow a = 2$$

Choice (4) is the answer.

2.47 Based on the information given in the problem, we have:

$$x = \sqrt{2} - 1 \tag{1}$$

$$f(x) = \cos x \tag{2}$$

$$gof(x) = 1 + \tan^2 x \tag{3}$$

Solving (2) and (3):

$$gof(x) = g(f(x)) = g(\cos x) = 1 + \tan^2 x \tag{4}$$

From trigonometry, we know that:

$$1 + \tan^2 x = \frac{1}{\cos^2 x} \tag{5}$$

Solving (4) and (5):

$$g(\cos x) = \frac{1}{\cos^2 x} \Rightarrow g(x) = \frac{1}{x^2} \tag{6}$$

$$\Rightarrow gog(x) = g(g(x)) = g\left(\frac{1}{x^2}\right) = \frac{1}{\left(\frac{1}{x^2}\right)^2} = x^4 \tag{7}$$

Solving (1) and (7):

$$(gog)\left(\sqrt{2} - 1\right) = \left(\sqrt{2} - 1\right)^4 = \left(2 - 2\sqrt{2} + 1\right)^2 = \left(3 - 2\sqrt{2}\right)^2 = 9 - 12\sqrt{2} + 8$$

$$\Rightarrow (gog)\left(\sqrt{2} - 1\right) = 17 - 12\sqrt{2}$$

Choice (4) is the answer.

2.48 Based on the information given in the problem, we have:

$$f(x) = \sqrt{\log\left(\frac{5x - x^2}{4}\right)}$$

The domain of a function in radical form, including even root, is determined by considering the radicand equal and greater than zero.

Moreover, the domain of a logarithmic function with the base of 10 can be determined by considering its argument greater than zero. Therefore:

$$\begin{cases} \log\left(\frac{5x - x^2}{4}\right) \geq 0 \\ \frac{5x - x^2}{4} > 0 \end{cases} \Rightarrow \begin{cases} \log\left(\frac{5x - x^2}{4}\right) \geq \log(1) \\ x^2 - 5x < 0 \end{cases} \Rightarrow \begin{cases} \frac{5x - x^2}{4} \geq 1 \\ x(x - 5) < 0 \end{cases} \Rightarrow \begin{cases} x^2 - 5x + 4 \leq 0 \\ x(x - 5) < 0 \end{cases}$$

$$\Rightarrow \begin{cases} (x - 4)(x - 1) \leq 0 \\ x(x - 5) < 0 \end{cases} \Rightarrow \begin{cases} 1 \leq x \leq 4 \\ 0 < x < 5 \end{cases} \overset{\cap}{\Rightarrow} 1 \leq x \leq 4$$

Choice (3) is the answer.

2.49 Based on the information given in the problem, we have:

$$f(x) = \frac{2x}{x + 2} \tag{1}$$

$$g(f(x)) = x \tag{2}$$

The problem can be solved as follows:

$$t \triangleq \frac{2x}{x+2} \tag{3}$$

$$\Rightarrow tx + 2t = 2x \Rightarrow 2x - tx = 2t \Rightarrow x(2-t) = 2t \Rightarrow x = \frac{2t}{2-t} \tag{4}$$

Solving (1) and (2):

$$g(f(x)) = g\left(\frac{2x}{x+2}\right) = x \tag{5}$$

Solving (3) and (5):

$$g(t) = x \tag{6}$$

Solving (4) and (6):

$$g(t) = \frac{2t}{2-t} \tag{7}$$

$$\Rightarrow g\left(\frac{1}{x}\right) = \frac{2\left(\frac{1}{x}\right)}{2 - \frac{1}{x}} = \frac{\frac{2}{x}}{\frac{2x-1}{x}} \tag{8}$$

$$\Rightarrow g\left(\frac{1}{x}\right) = \frac{2}{2x-1}$$

Choice (2) is the answer.

2.50 Based on the information given in the problem, we have:

$$f\left(\frac{1 - \cos 2x}{1 + \cos 2x}\right) = \cot x \tag{1}$$

From trigonometry, we know that:

$$1 - \cos 2x = 2\sin^2 x \tag{2}$$

$$1 + \cos 2x = 2\cos^2 x \tag{3}$$

$$\tan x = \frac{\sin x}{\cos x} \tag{4}$$

$$\cot x = \frac{1}{\tan x} \tag{5}$$

Therefore:

$$\frac{1 - \cos 2x}{1 + \cos 2x} = \frac{2\sin^2 x}{2\cos^2 x} = \tan^2 x \tag{6}$$

Solving (5) and (1):

$$f\left(\tan^2 x\right) = \cot x \tag{7}$$

Solving (7) and (5):

$$f\left(\tan^2 x\right) = \frac{1}{\tan x} \tag{8}$$

Defining a new parameter:

$$t \triangleq \tan^2 x \tag{9}$$

Solving (8) and (9):

$$g(t) = \frac{1}{\sqrt{t}} \Rightarrow g(x) = \frac{1}{\sqrt{x}}$$

Choice (2) is the answer.

2.51 Based on the information given in the problem, we have:

$$f(x) = \sqrt{\log_x(x^2 + 9)}$$

The domain of a function in radical form, including even root, is determined by considering the radicand equal and greater than zero.

In addition, the domain of a logarithmic function can be determined by considering its argument greater than zero. In addition, the base of the logarithm must be greater than zero but not equal to one. Therefore:

$$\begin{cases} \log_x\left(x^2 + 9\right) \geq 0 \\ x^2 + 9 > 0 \\ x > 0, x \neq 1 \end{cases} \Rightarrow \begin{cases} \log_x\left(x^2 + 9\right) \geq \log_x(1) \\ x^2 + 9 > 0 \\ x > 0, x \neq 1 \end{cases} \Rightarrow \begin{cases} x^2 + 9 \geq 1 \\ x^2 + 9 > 0 \\ x > 0, x \neq 1 \end{cases}$$

Note that $x^2 + 9 > 0$ and $x^2 + 8 \geq 0$ are true for any x. Hence:

$$\overset{\cap}{\Longrightarrow} x > 0, x \neq 1 \Rightarrow D_f = (0, \infty) - \{1\}$$

Choice (4) is the answer.

2.52 Based on the information given in the problem, we have:

$$f(x) = 2x - 2[x] + 1$$

Based on the definition, we know that:

$$[x] \leq x < [x] + 1 \xrightarrow{-[x]} 0 \leq x - [x] < 1 \xrightarrow{\times 2} 0 \leq 2x - 2[x] < 2 \xrightarrow{+1} 1 \leq 2x - 2[x] + 1 < 3$$

$$\Rightarrow 1 \leq f(x) < 3 \Rightarrow R_f = [1, 3)$$

Choice (2) is the answer.

2.53 Based on the information given in the problem, we have:

$$f(x) = x^2 + 1$$

$$g(x) = \sqrt{x-1}$$

$$\Rightarrow fog(x) = f(g(x)) = f\left(\sqrt{x-1}\right) = \left(\sqrt{x-1}\right)^2 + 1 = x - 1 + 1 = x$$

Next, we need to determine the domain of the function. As we know, the domain of a function in radical form, including even root, is determined by considering the radicand equal and greater than zero.

$$x - 1 \geq 0 \Rightarrow x \geq 1 \Rightarrow D_{fog} = [1, \infty)$$

Now, we can determine the range of the function based its domain as follows:

$$fog(x) = x \xrightarrow{\quad D_{fog} = [1, \infty) \quad} R_{fog} = [1, \infty)$$

Choice (2) is the answer.

2.54 Based on the information given in the problem, we have:

$$f(x) = \sqrt{x^2 - 2x + 3}$$

The problem can be solved as follows:

$$\Rightarrow f(x) = \sqrt{(x-1)^2 + 2}$$

As we know:

$$(x-1)^2 \geq 0 \xrightarrow{+2} (x-1)^2 + 2 \geq 2 \xrightarrow{\sqrt{\ }} \sqrt{(x-1)^2 + 2} \geq \sqrt{2} \Rightarrow f(x) \geq \sqrt{2} \Rightarrow R_{f(x)} = \left[\sqrt{2}, \infty\right)$$

Choice (1) is the answer.

2.55 Based on the information given in the problem, we have:

$$f(x) = |x - 2| \Rightarrow D_f = \mathbb{R}$$

Based on the definition, two functions are equivalent if they are equal, and their domains are the same.

Choice (1):

$$g_1(x) = \left|\frac{x^2 - 3x + 2}{x - 1}\right| = \left|\frac{(x-1)(x-2)}{x-1}\right| = |x - 2|$$

$$\Rightarrow D_{g_1} = \mathbb{R} - \{1\}$$

Therefore, the functions are not equivalent, since their domains are different. Note that $x = 1$ makes the denominator zero; thus, it is not in the domain.

Choice (2):

$$g_2(x) = \left| \frac{x^2 - 4}{x + 2} \right| = \left| \frac{(x-2)(x+2)}{x+2} \right| = |x - 2|$$

$$\Rightarrow D_{g_2} = \mathbb{R} - \{-2\}$$

Therefore, the functions are not equivalent because their domains are not the same. Note that $x = -2$ makes the denominator zero; thus, it must be excluded from the domain.

Choice (3):

$$g_3(x) = \frac{(x-2)^2}{|x-2|} = \frac{|x-2|^2}{|x-2|} = |x - 2|$$

$$\Rightarrow D_{g_3} = \mathbb{R} - \{-2\}$$

Therefore, the functions are not equivalent because their domains are not the same. Note that $x = -2$ makes the denominator zero; thus, it must be excluded from the domain.

Choice (4):

$$g_4(x) = \frac{|6x - 12|}{6} = \frac{6|x-2|}{6} = |x - 2|$$

$$\Rightarrow D_{g_4} = \mathbb{R}$$

Therefore, the functions are equivalent because their functions and domains are the same. Choice (4) is the answer.

2.56 Based on the information given in the problem, we have:

$$f(x) = \sqrt[3]{x} - \sqrt[3]{x + 2}$$

The line of $x = a$ is the axis of symmetry of the function of $f(x, y)$ if $f(a + x, y) = f(a - x, y)$.

For Choice (1), we have $x = -2 \Rightarrow a = -2$.

$$\begin{cases} f(-2 + x, y) = \sqrt[3]{(-2 + x)} - \sqrt[3]{(-2 + x) + 2} \\ f(-2 - x, y) = \sqrt[3]{(-2 - x)} - \sqrt[3]{(-2 - x) + 2} \end{cases}$$

$$\Rightarrow \sqrt[3]{x - 2} - \sqrt[3]{x} \neq \sqrt[3]{-x - 2} + \sqrt[3]{x}$$

$$\Rightarrow f(-2 + x, y) \neq f(-2 - x, y)$$

For Choice (2), we have $x = -1 \Rightarrow a = -1$.

$$\begin{cases} f(-1 + x, y) = \sqrt[3]{(-1 + x)} - \sqrt[3]{(-1 + x) + 2} \\ f(-1 - x, y) = \sqrt[3]{(-1 - x)} - \sqrt[3]{(-1 - x) + 2} \end{cases}$$

$$\Rightarrow \sqrt[3]{x-1} - \sqrt[3]{x+1} = -\sqrt[3]{x+1} + \sqrt[3]{x-1}$$

$$\Rightarrow f(-1+x, y) = f(-1-x, y)$$

For Choice (3), we have $x = 1 \Rightarrow a = 1$.

$$\begin{cases} f(1+x, y) = \sqrt[3]{(1+x)} - \sqrt[3]{(1+x)+2} \\ f(1-x, y) = \sqrt[3]{(1-x)} - \sqrt[3]{(1-x)+2} \end{cases}$$

$$\Rightarrow \sqrt[3]{x+1} - \sqrt[3]{x+3} \neq -\sqrt[3]{x-1} + \sqrt[3]{x-3}$$

$$\Rightarrow f(1+x, y) \neq f(1-x, y)$$

For Choice (4), we have $x = 2 \Rightarrow a = 2$.

$$\begin{cases} f(2+x, y) = \sqrt[3]{(2+x)} - \sqrt[3]{(2+x)+2} \\ f(2-x, y) = \sqrt[3]{(2-x)} - \sqrt[3]{(2-x)+2} \end{cases}$$

$$\Rightarrow \sqrt[3]{x+2} - \sqrt[3]{x+4} \neq -\sqrt[3]{x-2} + \sqrt[3]{x-4}$$

$$\Rightarrow f(2+x, y) \neq f(2-x, y)$$

In the calculations, the rule below was used because n was an odd number:

$$\sqrt[n]{-f(x)} = -\sqrt[n]{f(x)}$$

Choice (2) is the answer.

References

1. Rahmani-Andebili, M. (2021). Calculus – Practice Problems, Methods, and Solutions, Springer Nature, 2021.
2. Rahmani-Andebili, M. (2021). Precalculus – Practice Problems, Methods, and Solutions, Springer Nature, 2021.

Abstract

In this chapter, the basic and advanced problems of trigonometric equations and trigonometric identities are presented. The subjects include trigonometric equations, trigonometric identities, domain, range, period, half angle formulas, reciprocal identities, Pythagorean identities, expressing sum of sine and cosine as a product, expressing product of sine and cosine as a sum, even and odd functions, periodic functions, degrees to radians formula, cofunction formulas, unit circle, inverse trigonometric functions, and domain and range of inverse trigonometric functions. To help students study the chapter in the most efficient way, the problems are categorized in different levels based on their difficulty levels (easy, normal, and hard) and calculation amounts (small, normal, and large). Moreover, the problems are ordered from the easiest problem with the smallest computations to the most difficult problems with the largest calculations.

3.1. Calculate the value of $\sin x \cos x (1 - 2\sin^2 x)$ for $x = 7.5°$ [1, 2].

Difficulty level ● Easy ○ Normal ○ Hard
Calculation amount ● Small ○ Normal ○ Large

1) $\dfrac{\sqrt{3}}{8}$

2) $\dfrac{1}{8}$

3) $\dfrac{\sqrt{3}}{4}$

4) $\dfrac{1}{4}$

3.2. Calculate the value of $\tan^3 x + \cot^3 x$ if $\tan x + \cot x = 3$.

Difficulty level ● Easy ○ Normal ○ Hard
Calculation amount ● Small ○ Normal ○ Large

1) 18
2) 9
3) 27
4) 3

3.3. Which one of the following choices is correct about the extension of hyperbolic functions?

Difficulty level ● Easy ○ Normal ○ Hard
Calculation amount ● Small ○ Normal ○ Large

1)
$$\cosh(a + b) = \cosh a \cosh b - \sinh a \sinh b$$
$$\sinh(a + b) = \sinh a \cosh b + \cosh a \sinh b$$

2)
$$\cosh(a + b) = \cosh a \cosh b - \sinh a \sinh b$$
$$\sinh(a + b) = \sinh a \cosh b - \cosh a \sinh b$$

M. Rahmani-Andebili, *Calculus I*, https://doi.org/10.1007/978-3-031-45028-0_3

3) $\begin{aligned}\cosh(a+b) &= \cosh a \cosh b + \sinh a \sinh b \\ \sinh(a+b) &= \sinh a \cosh b - \cosh a \sinh b\end{aligned}$

4) $\begin{aligned}\cosh(a+b) &= \cosh a \cosh b + \sinh a \sinh b \\ \sinh(a+b) &= \sinh a \cosh b + \cosh a \sinh b\end{aligned}$

3.4. Calculate the value of $\tan(2\theta)$ if $\cot(\theta) = 5$.

Difficulty level   ● Easy   ○ Normal   ○ Hard
Calculation amount   ● Small   ○ Normal   ○ Large

1) $\dfrac{5}{12}$

2) $\dfrac{5}{13}$

3) $-\dfrac{5}{12}$

4) $-\dfrac{5}{13}$

3.5. Determine the value of $\tan(-2100°)$.

Difficulty level   ● Easy   ○ Normal   ○ Hard
Calculation amount   ● Small   ○ Normal   ○ Large

1) $\sqrt{3}$

2) $\dfrac{\sqrt{3}}{3}$

3) $-\sqrt{3}$

4) $-\dfrac{\sqrt{3}}{3}$

3.6. Simplify and calculate the final value of the following term:

$$\frac{1 + \cos(40°)}{\sin(40°)}$$

Difficulty level   ● Easy   ○ Normal   ○ Hard
Calculation amount   ● Small   ○ Normal   ○ Large

1) $\sin(20°)$
2) $\cos(20°)$
3) $\tan(20°)$
4) $\cot(20°)$

3.7. Determine the range of m if $\sin(\alpha) = \dfrac{3m-1}{4}$ and $\dfrac{\pi}{6} \le \alpha \le \dfrac{2\pi}{3}$.

Difficulty level   ● Easy   ○ Normal   ○ Hard
Calculation amount   ● Small   ○ Normal   ○ Large

1) $\left[1, \dfrac{2\sqrt{3}-1}{3}\right]$

2) $\left[1, \dfrac{2\sqrt{3}+1}{3}\right]$

3) $[1, 2]$

4) $\left[1, \dfrac{5}{3}\right]$

3.8. Determine the range of m if $\cos(x) = \dfrac{2m-1}{6}$ and $-\dfrac{\pi}{3} \le x \le \dfrac{\pi}{3}$.

Difficulty level   ● Easy   ○ Normal   ○ Hard
Calculation amount   ● Small   ○ Normal   ○ Large

1) $\left[2, \dfrac{7}{2}\right]$

2) $\left[\dfrac{3}{2}, \dfrac{7}{2}\right]$

3) $\left[2, \dfrac{5}{2}\right]$

4) $\left[\dfrac{3}{2}, \dfrac{5}{2}\right]$

3.9. What is the main period of $\cos^2(x) - 5\cos\left(\dfrac{2x}{3}\right)$?

Difficulty level   ● Easy   ○ Normal   ○ Hard
Calculation amount   ● Small   ○ Normal   ○ Large

1) π
2) 2π
3) 3π
4) 4π

3.10. What is the main period of $\sin^4\left(\dfrac{3x}{5}\right) + \cos^3\left(\dfrac{2x}{3}\right)$?

Difficulty level   ● Easy   ○ Normal   ○ Hard
Calculation amount   ● Small   ○ Normal   ○ Large

1) 3π
2) 5π
3) 15π
4) 30π

3.11. Determine the main period of $\sin^4\left(\dfrac{\pi x}{3}\right) + \cos(\pi x) + 5$.

Difficulty level   ● Easy   ○ Normal   ○ Hard
Calculation amount   ● Small   ○ Normal   ○ Large

1) 1
2) 2
3) 3
4) 6

3.12. Figure 3.1 illustrates part of the function of $y = \sin(kx)$. Determine the value of k.

Difficulty level   ● Easy   ○ Normal   ○ Hard
Calculation amount   ● Small   ○ Normal   ○ Large

1) $\dfrac{2}{3}$

2) $\dfrac{3}{4}$

3) $\dfrac{3}{2}$

4) $\dfrac{4}{3}$

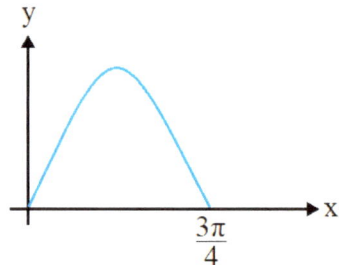

Figure 3.1 The graph of problem 3.12

3.13. Figure 3.2 illustrates part of the function of $y = \cos\left(\left(ax + \frac{1}{2}\right)\pi\right)$. Determine the value of a.

1) $\dfrac{1}{2}$

2) $\dfrac{3}{2}$

3) $\dfrac{2}{3}$

4) $\dfrac{7}{4}$

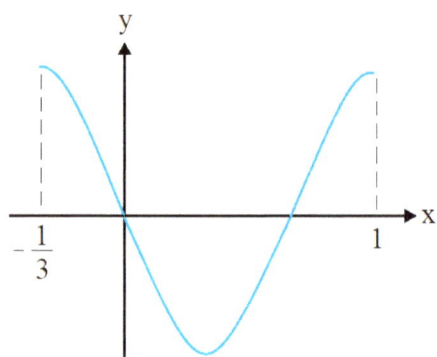

Figure 3.2 The graph of problem 3.13

3.14. Which one of the following choices is correct if $\alpha + \beta = 19\pi$?

1) $\sin(\alpha) = \sin(\beta)$
2) $\cos(\alpha) = \cos(\beta)$
3) $\tan(\alpha) = \tan(\beta)$
4) $\cot(\alpha) = \cot(\beta)$

3.15. Calculate the final value of the following equation.

$$\sin(5\pi + x) + \sin\left(x - \frac{\pi}{3}\right) + \sin\left(x + \frac{7\pi}{3}\right)$$

1) 0

2) $\sin\left(\dfrac{\pi}{3}\right)$

3) $2\sin\left(\dfrac{\pi}{3}\right)$

4) $-\sin\left(\dfrac{\pi}{3}\right)$

3.16. Calculate the value of the term below.

$$\sin\left[\arccos\left(\dfrac{-1}{2}\right) + \arcsin\left(\dfrac{-\sqrt{3}}{2}\right)\right]$$

Difficulty level   ○ Easy   ● Normal   ○ Hard
Calculation amount   ● Small   ○ Normal   ○ Large

1) $\dfrac{\sqrt{3}}{2}$

2) 1

3) $\dfrac{1}{2}$

4) -1

3.17. Calculate the value of $2\arctan\left(\dfrac{1}{2}\right)$.

Difficulty level   ○ Easy   ● Normal   ○ Hard
Calculation amount   ● Small   ○ Normal   ○ Large

1) $\arctan\dfrac{4}{3}$

2) $\arctan\dfrac{3}{4}$

3) $\arctan\dfrac{2}{3}$

4) $\arctan\dfrac{3}{2}$

3.18. What is the value of $\cos(\pi \sinh \ln 3)$?

Difficulty level   ○ Easy   ● Normal   ○ Hard
Calculation amount   ● Small   ○ Normal   ○ Large

1) -1

2) $-\dfrac{1}{2}$

3) $\dfrac{\sqrt{2}}{2}$

4) $\dfrac{\sqrt{3}}{2}$

3.19. Calculate the value of $\cos(20°)$ if $\sin(50°) + \sin(10°) = m$.

Difficulty level   ○ Easy   ● Normal   ○ Hard
Calculation amount   ● Small   ○ Normal   ○ Large

1) $\dfrac{m}{2}$

2) m

3) $2m$

4) $\dfrac{2m}{3}$

3.20. Simplify and calculate the final value of the following term:

$$\frac{(1 + \tan^2(5°)) \sin(10°)}{(1 - \tan^2(5°)) \tan(10°)}$$

Difficulty level   ○ Easy   ● Normal   ○ Hard
Calculation amount   ● Small   ○ Normal   ○ Large
1) $\tan(15°)$
2) $\tan(25°)$
3) $\tan(35°)$
4) $\tan(45°)$

3.21. Which one of the following relations is correct if $\cot(\alpha) = m$ and $\cos(\alpha) = n$?
Difficulty level   ○ Easy   ● Normal   ○ Hard
Calculation amount   ● Small   ○ Normal   ○ Large
1) $m^2(1 + n^2) = n^2$
2) $m^2(1 - n^2) = n^2$
3) $m^2(2 + n^2) = 1$
4) $m^2(2 - n^2) = 1$

3.22. Determine the main period of $\sin(3x) \cos(5x) + 11$.
Difficulty level   ○ Easy   ● Normal   ○ Hard
Calculation amount   ● Small   ○ Normal   ○ Large
1) π
2) 2π
3) $\frac{2\pi}{3}$
4) $\frac{2\pi}{5}$

3.23. Calculate the value of $\text{arc}\left(\cos\left(\sin\left(\frac{4\pi}{3}\right)\right)\right)$.
Difficulty level   ○ Easy   ● Normal   ○ Hard
Calculation amount   ● Small   ○ Normal   ○ Large
1) $\frac{\pi}{6}$
2) $\frac{5\pi}{6}$
3) $\frac{\pi}{3}$
4) $-\frac{\pi}{6}$

3.24. Calculate the value of $\text{arc}\left(\sin\left(\sin\left(\frac{17\pi}{5}\right)\right)\right)$.
Difficulty level   ○ Easy   ● Normal   ○ Hard
Calculation amount   ● Small   ○ Normal   ○ Large
1) $\frac{2\pi}{5}$
2) $\frac{3\pi}{5}$
3) $-\frac{2\pi}{5}$
4) $-\frac{3\pi}{5}$

3.25. Calculate the value of $\arc\left(\cos\left(\cos\left(\frac{19\pi}{5}\right)\right)\right)$.

1) $\frac{\pi}{5}$

2) $\frac{4\pi}{5}$

3) $-\frac{\pi}{5}$

4) $-\frac{4\pi}{5}$

3.26. Calculate the value of $\tan\left(2\arc\left(\tan\left(\frac{1}{2}\right)\right)\right)$.

1) 1

2) $\frac{3}{4}$

3) $\frac{4}{3}$

4) $\frac{3}{5}$

3.27. Calculate the final value of $\sin\left(\arc\left(\sin\left(\frac{3}{5}\right)\right) + \arc\left(\tan\left(\frac{3}{4}\right)\right)\right)$.

1) $\frac{10}{13}$

2) $\frac{9}{13}$

3) $\frac{12}{35}$

4) $\frac{24}{25}$

3.28. Calculate the final value of $\arc\left(\cot\left(-\frac{4}{3}\right)\right) - \arc\left(\cot\left(\frac{3}{4}\right)\right)$.

1) π

2) $\frac{2\pi}{3}$

3) $\frac{\pi}{2}$

4) $\frac{\pi}{3}$

3.29. Calculate the final value of $\arc(\tan(5)) + \arc\left(\tan\left(\frac{3}{2}\right)\right)$.

1) $\frac{\pi}{4}$

2) $-\frac{\pi}{4}$

3) $\dfrac{3\pi}{4}$

4) $\dfrac{5\pi}{4}$

3.30. Calculate the final value of $\sin\left(\arc\left(\cos\left(\dfrac{3}{5}\right)\right)\right) + \cos\left(\arc\left(\sin\left(-\dfrac{4}{5}\right)\right)\right)$.

     Difficulty level   ○ Easy   ● Normal   ○ Hard

     Calculation amount   ● Small   ○ Normal   ○ Large

1) $\dfrac{7}{5}$

2) $-\dfrac{1}{5}$

3) $\dfrac{1}{5}$

4) $-\dfrac{7}{5}$

3.31. Figure 3.3 shows a unit circle. Which one of the choices shows the value of $\tan(\theta)$ and $\cot(\theta)$, respectively?

     Difficulty level   ○ Easy   ● Normal   ○ Hard

     Calculation amount   ● Small   ○ Normal   ○ Large

1) *OA, OB*

2) *HA, HB*

3) *OA, AB*

4) *OB, BH*

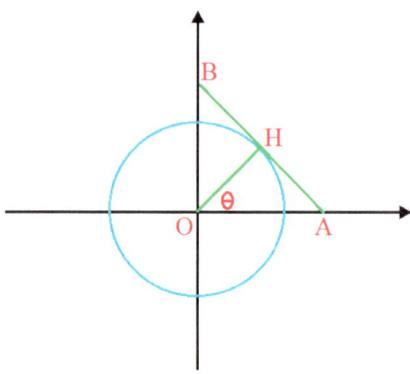

Figure 3.3 The graph of problem 3.31

3.32. Figure 3.4 illustrates a unit circle. Which one of the choices shows the value of $\sec(\theta)$?

     Difficulty level   ○ Easy   ● Normal   ○ Hard

     Calculation amount   ● Small   ○ Normal   ○ Large

1) *HA*

2) *MB*

3) *OB*

4) *OM*

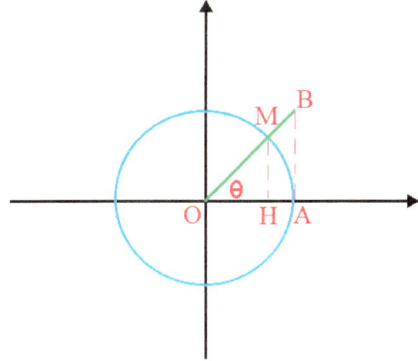

Figure 3.4 The graph of problem 3.32

3.33. Figure 3.5 illustrates a unit circle. Which one of the choices shows the value of csc(θ)?

Difficulty level   ○ Easy   ● Normal   ○ Hard
Calculation amount   ● Small   ○ Normal   ○ Large
1) *MB*
2) *OB*
3) *HC*
4) *OM*

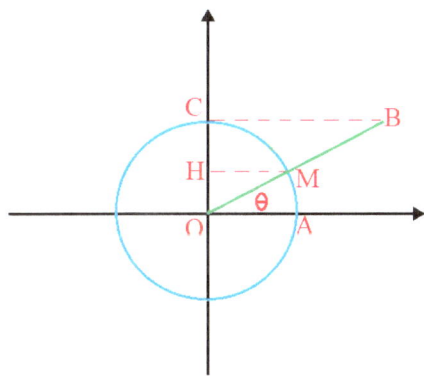

Figure 3.5 The graph of problem 3.33

3.34. Calculate the value of $\mathrm{arc}\left(\tan\left(\frac{2}{3}\right)\right) + \mathrm{arc}\left(\tan\left(\frac{1}{5}\right)\right)$.

Difficulty level   ○ Easy   ● Normal   ○ Hard
Calculation amount   ● Small   ○ Normal   ○ Large
1) $\frac{\pi}{6}$
2) $\frac{\pi}{4}$
3) $\frac{\pi}{3}$
4) $\frac{\pi}{2}$

3.35. Calculate the final value of the term below.

$$\mathrm{arc}(\tan(m)) + \mathrm{arc}\left(\tan\left(\frac{1}{m}\right)\right) + \mathrm{arc}(\cot(m)) + \mathrm{arc}(\cot(-m))$$

1) π or 2π

2) $\dfrac{\pi}{2}$ or $\dfrac{3\pi}{2}$

3) $\dfrac{3\pi}{2}$

4) $\dfrac{\pi}{2}$

3.36. Determine the range of x in the inequality below. Herein, x is an acute angle.

$$-1 \le \cos(4x)\cos(2x) + \sin(4x)\sin(2x) \le 0$$

1) $\left[\dfrac{\pi}{6}, \dfrac{3\pi}{8}\right]$

2) $\left[\dfrac{\pi}{8}, \dfrac{\pi}{4}\right]$

3) $\left[\dfrac{\pi}{6}, \dfrac{\pi}{3}\right]$

4) $\left[\dfrac{\pi}{4}, \dfrac{\pi}{2}\right]$

3.37. Calculate the value of $\tan(2y)$ if $\tan(x + y) = 5$ and $\tan(x - y) = 7$.

1) $\dfrac{1}{18}$

2) $-\dfrac{1}{18}$

3) $\dfrac{1}{36}$

4) $-\dfrac{1}{36}$

3.38. Simplify and calculate the value of the following term:

$$\frac{\sin\left(\dfrac{5\pi}{12}\right) + \cos\left(\dfrac{5\pi}{12}\right)}{\sin\left(\dfrac{5\pi}{12}\right) - \cos\left(\dfrac{5\pi}{12}\right)}$$

1) $\sqrt{3}$

2) $\dfrac{\sqrt{3}}{3}$

3) $-2\sqrt{3}$

4) $-\dfrac{\sqrt{3}}{3}$

3.39. Figure 3.6 illustrates part of the function of $y = a \sin(b\pi x)$. Determine the value of $a + b$.

1) $\dfrac{4}{3}$

2) $\dfrac{5}{3}$

3) $\dfrac{7}{3}$

4) $\dfrac{8}{3}$

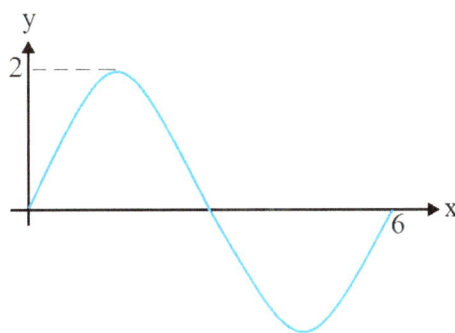

Figure 3.6 The graph of problem 3.39

3.40. Figure 3.7 illustrates part of the function of $y = a \sin(b\pi x)$. Determine the value of $a \times b$.

1) -6

2) -3

3) $\dfrac{9}{2}$

4) 6

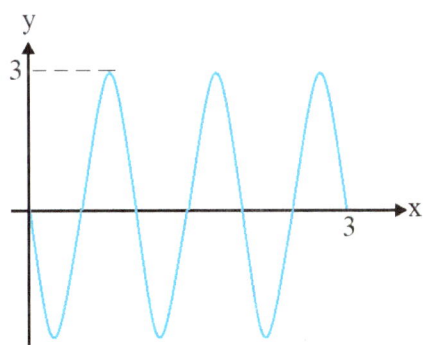

Figure 3.7 The graph of problem 3.40

3.41. Figure 3.8 illustrates part of the function of $y = a \sin\left(\left(bx + \dfrac{1}{2}\right)\pi\right)$. Determine the value of $a \times b$.

1) 2

2) $\dfrac{5}{2}$

3) 3

4) $\frac{7}{2}$

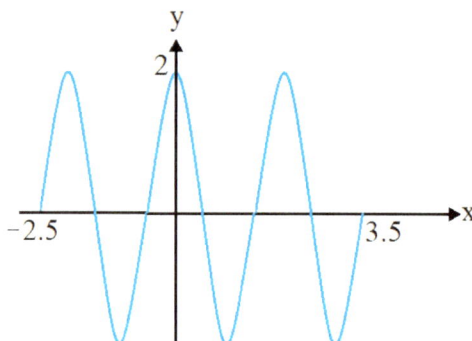

Figure 3.8 The graph of problem 3.41

3.42. Simplify and calculate the value of the following term:

$$\frac{\cos(5^\circ)\cos(10^\circ)\cos(20^\circ)}{\cos(50^\circ)}$$

Difficulty level   ○ Easy   ● Normal   ○ Hard
Calculation amount   ○ Small   ● Normal   ○ Large

1) $\dfrac{1}{4\cos(85^\circ)}$

2) $\dfrac{1}{8\cos(85^\circ)}$

3) $\dfrac{1}{8\sin(85^\circ)}$

4) $\dfrac{1}{4\sin(85^\circ)}$

3.43. Calculate the value of $\tan\left(\dfrac{x}{2}\right)$ if $\sin(x)+\cos(x)=\dfrac{7}{5}$.

Difficulty level   ○ Easy   ● Normal   ○ Hard
Calculation amount   ○ Small   ● Normal   ○ Large

1) 2 or 3

2) $\dfrac{1}{2}$ or $\dfrac{1}{3}$

3) 2 or $\dfrac{3}{5}$

4) 3 or $\dfrac{2}{5}$

3.44. Simplify and calculate the value of the following term:

$$\frac{\sin^4(\alpha)-\cos^4(\alpha)}{\sin(\alpha)\cos(\alpha)}$$

Difficulty level   ○ Easy   ● Normal   ○ Hard
Calculation amount   ○ Small   ● Normal   ○ Large

1) $2 \cot(2\alpha)$

2) $-2 \cot(2\alpha)$

3) $2 \tan(3\alpha)$

4) $-2 \tan(3\alpha)$

3.45. Calculate the value of $\cot^2(2\alpha)$ if $\sin^4(\alpha) + \cos^4(\alpha) = \dfrac{1}{2}$.

Difficulty level   ○ Easy   ● Normal   ○ Hard

Calculation amount   ○ Small   ● Normal   ○ Large

1) 1

2) 2

3) 0

4) 3

3.46. Calculate the value of the following relation for $x = \dfrac{3\pi}{8}$:

$$\sin^3(x)\cos(x) - \cos^3(x)\sin(x) + 3\sin^2(x)\cos^2(x)$$

Difficulty level   ○ Easy   ● Normal   ○ Hard

Calculation amount   ○ Small   ● Normal   ○ Large

1) $\dfrac{3}{8}$

2) $\dfrac{5}{8}$

3) $-\dfrac{5}{8}$

4) $-\dfrac{3}{8}$

3.47. Calculate the value of the following relation for $\alpha = \dfrac{\pi}{15}$:

$$\frac{\sin(2\alpha) + \sin(5\alpha) + \sin(8\alpha)}{\cos(2\alpha) + \cos(5\alpha) + \cos(8\alpha)}$$

Difficulty level   ○ Easy   ● Normal   ○ Hard

Calculation amount   ○ Small   ● Normal   ○ Large

1) $-\dfrac{\sqrt{3}}{3}$

2) $-\sqrt{3}$

3) $\sqrt{3}$

4) $\dfrac{\sqrt{3}}{3}$

3.48. Calculate the value of the following relation for $x = \dfrac{\pi}{12}$:

$$(\sin(x) - \cos(x) + 2)(\sin(x) - \cos(x) - 2)$$

Difficulty level   ○ Easy   ● Normal   ○ Hard

Calculation amount   ○ Small   ● Normal   ○ Large

1) $\dfrac{7}{2}$

2) $\dfrac{5}{2}$

3) $-\dfrac{5}{2}$

4) $-\dfrac{7}{2}$

3.49. Calculate the value of $4\sin^2(\alpha)\cos^2(\alpha)(\tan(\alpha) + \cot(\alpha))^2$.

Difficulty level   ○ Easy   ● Normal   ○ Hard
Calculation amount   ○ Small   ● Normal   ○ Large
1) 1
2) 2
3) 3
4) 4

3.50. Determine the number of roots of the equation below.

$$\sin(\pi x)\cos^2(\pi x) + \sin^2(\pi x)\cos(\pi x) = 0.$$

Difficulty level   ○ Easy   ● Normal   ○ Hard
Calculation amount   ○ Small   ● Normal   ○ Large
1) 11
2) 12
3) 13
4) 14

3.51. Figure 3.9 illustrates part of the function of $y = \dfrac{1}{2} + 2\cos(mx)$. Determine the value of the function for $x = \dfrac{16\pi}{3}$.

Difficulty level   ○ Easy   ● Normal   ○ Hard
Calculation amount   ○ Small   ● Normal   ○ Large
1) $-\dfrac{1}{2}$

2) $\dfrac{1}{2}$

3) 1
4) 0

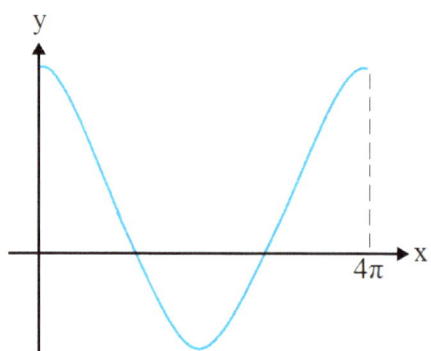

Figure 3.9 The graph of problem 3.51

3.52. Figure 3.10 shows part of the function of $y = 1 + \sin(mx)$. Determine the value of the function for $x = \frac{7\pi}{6}$.

1) 0

2) $\frac{1}{2}$

3) 1

4) 2

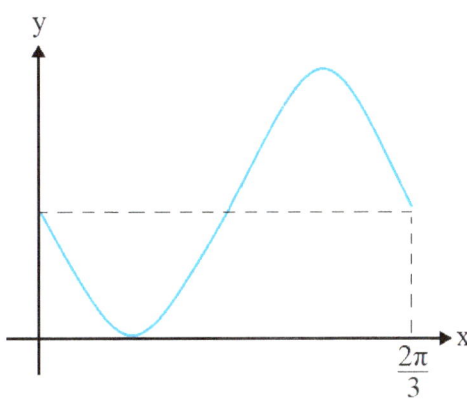

Figure 3.10 The graph of problem 3.52

3.53. Figure 3.11 shows part of the function of $y = a - \sin(b\pi x)$. Determine the value of the function for $x = \frac{25}{3}$.

1) 2

2) $\frac{5}{2}$

3) 3

4) $\frac{7}{2}$

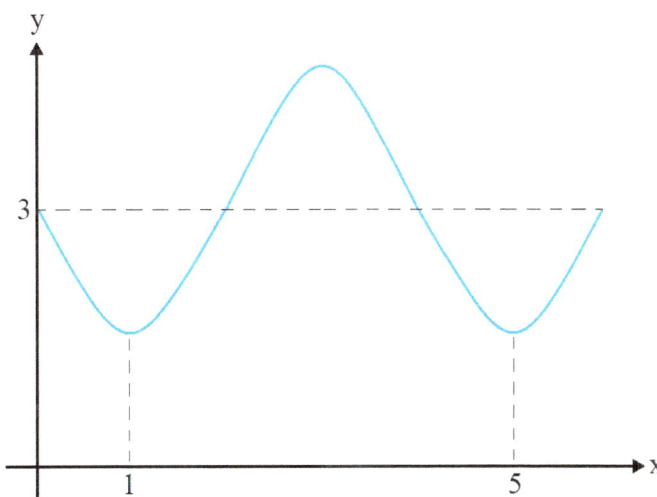

Figure 3.11 The graph of problem 3.53

3.54. Figure 3.12 shows the function of $y = a + b \cos\left(\frac{\pi}{2} x\right)$ for $0 < x < 4$. Determine the value of b.

Difficulty level   ○ Easy   ● Normal   ○ Hard
Calculation amount   ○ Small   ● Normal   ○ Large
1) -2
2) -1
3) 1
4) 2

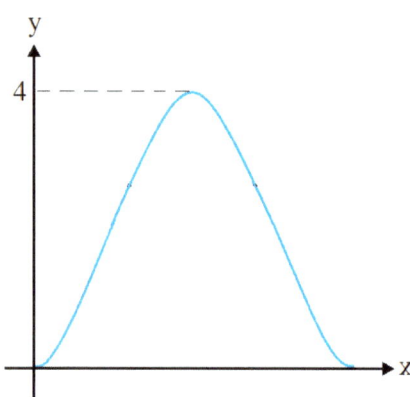

Figure 3.12   The graph of problem 3.54

3.55. Figure 3.13 shows the function of $y = 1 + a \sin(b\pi x)$ for $0 < x < \frac{4}{3}$. Determine the value of $a + b$.

Difficulty level   ○ Easy   ● Normal   ○ Hard
Calculation amount   ○ Small   ● Normal   ○ Large
1) 3
2) 4
3) 5
4) 6

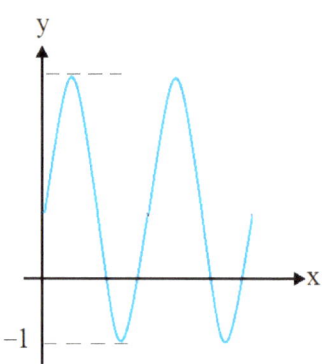

Figure 3.13   The graph of problem 3.55

3.56. Figure 3.14 shows part the function of $y = a - 2 \cos\left(bx + \frac{\pi}{2}\right)$. Determine the value of $a + b$.

Difficulty level   ○ Easy   ● Normal   ○ Hard
Calculation amount   ○ Small   ● Normal   ○ Large
1) $\frac{1}{2}$
2) 1
3) $\frac{3}{2}$
4) 2

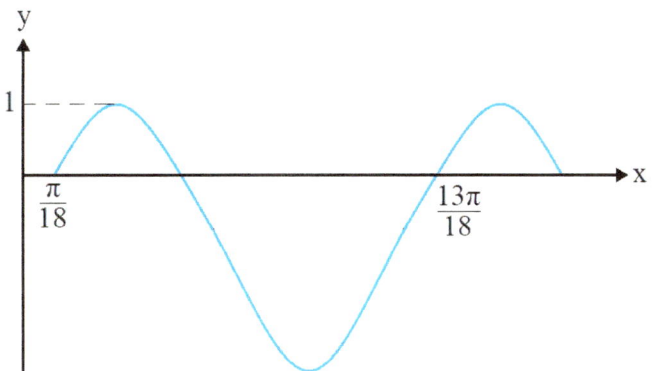

Figure 3.14 The graph of problem 3.56

3.57. Calculate the value of $\cos(25° - \alpha)$ if $\tan(\alpha + 20°) = \frac{3}{4}$.

1) 5
2) 6
3) 7
4) 8

3.58. Calculate the value of $\tan\left(\frac{\pi}{4} + \alpha\right)$ assuming that α is an acute angle and $\sin(\alpha) = \frac{3}{5}$.

1) -7
2) $-\frac{1}{7}$
3) $\frac{1}{7}$
4) 7

3.59. Calculate the value of $\tan\left(\frac{\pi}{4} - \alpha\right)$ if $\tan\left(\frac{\pi}{2} - \alpha\right) = \frac{2}{3}$.

1) $-\frac{1}{3}$
2) $-\frac{1}{5}$
3) $\frac{1}{5}$
4) $\frac{1}{3}$

3.60. Calculate the value of $\tan(2a)$ while we know that $\tan(a + b) = \frac{2}{5}$ and $\tan(a - b) = \frac{3}{7}$.

1) $-\frac{1}{3}$
2) $-\frac{1}{2}$
3) 3
4) 1

3.61. Calculate the value of tan(x) if we have:

$$\frac{\sin\left(x - \frac{\pi}{4}\right)}{\cos\left(x - \frac{\pi}{4}\right)} = 2$$

Difficulty level   ○ Easy   ● Normal   ○ Hard
Calculation amount   ○ Small   ● Normal   ○ Large
1) -3
2) $\dfrac{1}{3}$
3) $\dfrac{2}{3}$
4) 3

3.62. Calculate the value of $(1 + \tan(\alpha))(1 + \tan(\beta))$ if $\alpha + \beta = \dfrac{\pi}{4}$.

Difficulty level   ○ Easy   ● Normal   ○ Hard
Calculation amount   ○ Small   ● Normal   ○ Large
1) -2
2) 2
3) $\dfrac{1}{3}$
4) $-\dfrac{1}{2}$

3.63. Calculate the value of $\tan\left(\dfrac{\pi}{4} + \alpha\right) - \tan\left(\dfrac{\pi}{4} - \alpha\right)$.

Difficulty level   ○ Easy   ● Normal   ○ Hard
Calculation amount   ○ Small   ● Normal   ○ Large
1) $2\tan(2\alpha)$
2) $2\cos(2\alpha)$
3) 0
4) $2\sin(2\alpha)$

3.64. Calculate the value of tan(2α) if $\tan\left(\dfrac{\pi}{4} - \alpha\right) = \dfrac{1}{5}$.

Difficulty level   ○ Easy   ● Normal   ○ Hard
Calculation amount   ○ Small   ● Normal   ○ Large
1) 1.5
2) 1.8
3) 2.4
4) 2.5

3.65. Calculate the value of tan($2\alpha - \beta$) if $\tan(\alpha) = 2$ and $\tan(\beta) = \dfrac{1}{3}$.

Difficulty level   ○ Easy   ● Normal   ○ Hard
Calculation amount   ○ Small   ● Normal   ○ Large
1) -3
2) -2
3) 0.5
4) 3

3.66. Determine the common solution of the equation of $\cos(3x) + \cos(x) = 0$ assuming $\cos(x) \neq 0$.

Difficulty level   ○ Easy   ● Normal   ○ Hard
Calculation amount   ○ Small   ● Normal   ○ Large

1) $\dfrac{k\pi}{2} + \dfrac{\pi}{4}$

2) $\dfrac{k\pi}{2} + \dfrac{\pi}{8}$

3) $k\pi - \dfrac{\pi}{4}$

4) $k\pi + \dfrac{\pi}{4}$

3.67. Calculate the sum of the positive acute roots of the equation of $\tan(4x) = \cot(x)$.

Difficulty level   ○ Easy   ● Normal   ○ Hard
Calculation amount   ○ Small   ● Normal   ○ Large

1) $\dfrac{2\pi}{5}$

2) $\dfrac{4\pi}{5}$

3) $\dfrac{3\pi}{5}$

4) $\dfrac{\pi}{5}$

3.68. Determine the common solution of the equation of $2\sin^2(x) + 3\cos(x) = 0$.

Difficulty level   ○ Easy   ● Normal   ○ Hard
Calculation amount   ○ Small   ● Normal   ○ Large

1) $2k\pi \pm \dfrac{2\pi}{3}$

2) $2k\pi \pm \dfrac{\pi}{3}$

3) $2k\pi \pm \dfrac{5\pi}{6}$

4) $k\pi - \dfrac{\pi}{3}$

3.69. Determine the common solution of the equation of $2\sin^2(x) = 3\cos(x)$.

Difficulty level   ○ Easy   ● Normal   ○ Hard
Calculation amount   ○ Small   ● Normal   ○ Large

1) $k\pi \pm \dfrac{\pi}{6}$

2) $k\pi \pm \dfrac{\pi}{3}$

3) $2k\pi \pm \dfrac{\pi}{6}$

4) $2k\pi \pm \dfrac{\pi}{3}$

3.70. Two lines with the equations of $x\tan(\alpha) + y\cot(\alpha) = 2$ and $x\tan(\alpha) - y\cot(\alpha) = 1$ are intersecting each other at point M. By changing the value of α, what is the position equation of the point?

Difficulty level   ○ Easy   ○ Normal   ● Hard
Calculation amount   ● Small   ○ Normal   ○ Large

1) $y = \dfrac{1}{x}$

2) $y = \dfrac{3}{x}$

3) $y = \dfrac{1}{4x}$

4) $y = \dfrac{3}{4x}$

3.71. What is the position equation of the point of $(2 - 3\sin(\alpha), 1 + 4\cos(\alpha))$ if the value of α changes?

1) Circle
2) Ellipse
3) Parabola
4) Hyperbola

3.72. What is the position equation of the point of $(2 - 5\cos(\alpha), 4)$ if the value of α changes?

1) A horizontal line
2) A vertical line
3) A horizontal line segment
4) A vertical line segment

3.73. Calculate the value of y if $2\cos(x - y) + 3\sin(x + y) = 5$ and $0 < x, y < 2\pi$.

1) $\dfrac{\pi}{3}$ or $\dfrac{2\pi}{3}$
2) $\dfrac{\pi}{4}$ or $\dfrac{5\pi}{4}$
3) $\dfrac{\pi}{6}$ or $\dfrac{5\pi}{6}$
4) $\dfrac{\pi}{2}$ or $\dfrac{3\pi}{2}$

3.74. Calculate the value of m if $\tan(\alpha) \neq \tan(\beta)$, $\alpha + \beta = \dfrac{\pi}{4}$, and α and β are the two roots of the equation below.

$$\tan^2(x) + (m + 2)\tan(x) + 2m - 2 = 0$$

1) 1
2) 3
3) 5
4) 7

3.75. Calculate the final value of the following relation:

$$\frac{\sin^6(\alpha) + \cos^6(\alpha) + 3\sin^2(\alpha)\cos^2(\alpha)}{\sin^4(\alpha) + \cos^4(\alpha) + 2\sin^2(\alpha)\cos^2(\alpha)}$$

1) $\sin^2(\alpha)$
2) $\cos^2(\alpha)$
3) $\sin^2(\alpha) - \cos^2(\alpha)$
4) 1

3.76. Calculate the final value of the relation below.

$$\frac{\sin(135°)\cos(210°) + \cos(135°)\sin(420°)}{\tan(210°)\cot(420°) + \cot(120°)\tan(330°)}$$

Difficulty level   ○ Easy   ○ Normal   ● Hard
Calculation amount   ○ Small   ● Normal   ○ Large

1) $-\dfrac{\sqrt{6}}{4}$

2) $-\dfrac{3\sqrt{6}}{4}$

3) $-\dfrac{\sqrt{6}}{2}$

4) $-\dfrac{3\sqrt{6}}{2}$

3.77. Calculate the final value of $(1 + \cot(x))(1 + \cot(y))$ if $x + y = k\pi + \dfrac{\pi}{4}$.

Difficulty level   ○ Easy   ○ Normal   ● Hard
Calculation amount   ○ Small   ● Normal   ○ Large
1) $\tan(x)\tan(y)$
2) $2\tan(x)\tan(y)$
3) $\cot(x)\cot(y)$
4) $2\cot(x)\cot(y)$

3.78. Determine the common solution of the equation below.

$$(\sin(x) - \tan(x))\tan\left(\frac{3\pi}{2} - x\right) = \cos\left(\frac{4\pi}{3}\right)$$

Difficulty level   ○ Easy   ○ Normal   ● Hard
Calculation amount   ○ Small   ● Normal   ○ Large

1) $k\pi - \dfrac{\pi}{6}$

2) $k\pi + \dfrac{\pi}{3}$

3) $2k\pi \pm \dfrac{\pi}{3}$

4) $2k\pi \pm \dfrac{\pi}{6}$

3.79. Calculate the sum of the roots of the equation below for $x \in [0, \pi]$.

$$\sin(2x)(\sin(x) + \cos(x)) = \cos(2x)(\cos(x) - \sin(x))$$

Difficulty level   ○ Easy   ○ Normal   ● Hard
Calculation amount   ○ Small   ● Normal   ○ Large

1) $\dfrac{3\pi}{4}$

2) $\dfrac{5\pi}{4}$

3) $\dfrac{3\pi}{2}$

4) $\dfrac{7\pi}{4}$

3.80. Determine the common solution of the equation of $\sqrt{2}\sin\left(\dfrac{\pi}{4}-x\right)=1+\sin\left(\dfrac{5\pi}{2}+x\right)$.

Difficulty level   ○ Easy   ○ Normal   ● Hard
Calculation amount   ○ Small   ● Normal   ○ Large

1) $k\pi+\dfrac{\pi}{2}$

2) $2k\pi-\dfrac{\pi}{4}$

3) $2k\pi-\dfrac{\pi}{2}$

4) $2k\pi+\dfrac{\pi}{2}$

3.81. Which one of the following choices shows one of the common solutions of the equation of $\cos(2x)+\sqrt{3}\sin(2x)=1$?

Difficulty level   ○ Easy   ○ Normal   ● Hard
Calculation amount   ○ Small   ● Normal   ○ Large

1) $k\pi-\dfrac{\pi}{6}$

2) $k\pi-\dfrac{\pi}{3}$

3) $k\pi+\dfrac{\pi}{6}$

4) $k\pi+\dfrac{\pi}{3}$

3.82. Calculate the value of the following term:

$$\frac{\cos 3\alpha+\sin\alpha\sin 2\alpha}{\sin 3\alpha-\sin 2\alpha\cos\alpha}\times\frac{\sin\alpha}{\cos\alpha}$$

Difficulty level   ○ Easy   ○ Normal   ● Hard
Calculation amount   ○ Small   ● Normal   ○ Large

1) $\tan\alpha$

2) $\cot\alpha$

3) 1

4) -1

References

1. Rahmani-Andebili, M. (2021). Calculus – Practice Problems, Methods, and Solutions, Springer Nature, 2021.
2. Rahmani-Andebili, M. (2021). Precalculus – Practice Problems, Methods, and Solutions, Springer Nature, 2021.

Abstract

In this chapter, the problems of the third chapter are fully solved, in detail, step-by-step, and with different methods.

4.1. From trigonometry, we know that [1, 2]:

$$\sin 2x = 2 \sin(x) \cos(x)$$

$$1 - \cos 2x = 2 \sin^2 x$$

$$\sin 30^\circ = \frac{1}{2}$$

Therefore:

$$\sin x \cos x \left(1 - 2 \sin^2 x\right) = \left(\frac{1}{2} \sin 2x\right)(\cos 2x) = \frac{1}{4} \sin 4x$$

For $x = 7.5^\circ$, we have:

$$\frac{1}{4} \times \sin(4 \times 7.5^\circ) = \frac{1}{4} \sin 30^\circ = \frac{1}{8}$$

Choice (2) is the answer.

4.2. Based on the information given in the problem, we have:

$$\tan x + \cot x = 3$$

From algebra, we know that:

$$a^3 + b^3 = (a + b)^3 - 3ab(a + b)$$

In addition, from trigonometry, we know that:

$$\tan x \cot x = 1$$

Therefore:

$$\tan^3 x + \cot^3 x = (\tan x + \cot x)^3 - 3\tan x \cot x(\tan x + \cot x) = (3)^3 - 3(1)(3)$$

$$\Rightarrow \tan^3 x + \cot^3 x = 18$$

Choice (1) is the answer.

4.3. From trigonometry of hyperbolic functions, we know that:

$$\cosh(a \pm b) = \cosh a \cosh b \pm \sinh a \sinh b$$
$$\sinh(a \pm b) = \sinh a \cosh b \pm \cosh a \sinh b$$

Choice (4) is the answer.

4.4. From trigonometry, we know that:

$$\tan(\theta) = \frac{1}{\cot(\theta)}$$

$$\tan(2\theta) = \frac{2\tan(\theta)}{1 - \tan^2(\theta)}$$

Based on the information given in the problem:

$$\cot(\theta) = 5 \Rightarrow \tan(\theta) = \frac{1}{5}$$

Therefore:

$$\tan(2\theta) = \frac{2\tan(\theta)}{1 - \tan^2(\theta)} = \frac{2 \times \frac{1}{5}}{1 - \left(\frac{1}{5}\right)^2} = \frac{\frac{2}{5}}{\frac{24}{25}}$$

$$\Rightarrow \tan(2\theta) = \frac{5}{12}$$

Choice (1) is the answer.

4.5. From trigonometry, we know that:

$$\tan(\alpha + n\pi) = \tan(\alpha), \forall n \in \mathbb{Z}$$

$$\tan(-\alpha) = -\tan(\alpha)$$

$$\tan(60^\circ) = \sqrt{3}$$

Therefore:

$$\tan(-2100^\circ) = -\tan(2100^\circ) = -\tan(12 \times 180 - 60^\circ) = -\tan(-60^\circ) = \tan(60^\circ)$$

$$\Rightarrow \tan(-2100^\circ) = \sqrt{3}$$

Choice (1) is the answer.

4.6. From trigonometry, we know that:

$$1 + \cos(\theta) = 2\cos^2\left(\frac{\theta}{2}\right)$$

$$\sin(\theta) = 2\sin\left(\frac{\theta}{2}\right)\cos\left(\frac{\theta}{2}\right)$$

$$\cot(\theta) = \frac{\cos(\theta)}{\sin(\theta)}$$

Therefore:

$$\frac{1 + \cos(40°)}{\sin(40°)} = \frac{2\cos^2(20°)}{2\sin(20°)\cos(20°)} = \frac{\cos(20°)}{\sin(20°)}$$

$$\Rightarrow \frac{1 + \cos(40°)}{\sin(40°)} = \cot(20°)$$

Choice (4) is the answer.

4.7. For the given range of α, we can conclude that:

$$\frac{\pi}{6} \leq \alpha \leq \frac{2\pi}{3} \Rightarrow \frac{1}{2} \leq \sin(\alpha) \leq 1$$

Therefore, based on the given information, i.e., $\sin(\alpha) = \frac{3m-1}{4}$, we can write:

$$\frac{1}{2} \leq \frac{3m-1}{4} \leq 1 \Rightarrow 2 \leq 3m - 1 \leq 4 \Rightarrow 3 \leq 3m \leq 5$$

$$\Rightarrow 1 \leq m \leq \frac{5}{3}$$

Choice (4) is the answer.

4.8. For the given range of x, we can conclude that:

$$-\frac{\pi}{3} \leq x \leq \frac{\pi}{3} \Rightarrow \frac{1}{2} \leq \cos(x) \leq 1$$

Therefore, based on the given information, i.e., $\cos(x) = \frac{2m-1}{6}$, we can write:

$$\frac{1}{2} \leq \frac{2m-1}{6} \leq 1 \Rightarrow 3 \leq 2m - 1 \leq 6 \Rightarrow 4 \leq 2m \leq 7$$

$$\Rightarrow 2 \leq m \leq \frac{7}{2}$$

Choice (1) is the answer.

4.9. From trigonometry, we know that:

$$f_1(x) = \cos^{2n}(ax), \forall n \in \mathbb{Z} \Rightarrow T_1 = \frac{\pi}{|a|}$$

$$f_2(x) = \cos^{2n+1}(ax), \forall n \in \mathbb{Z} \Rightarrow T_2 = \frac{2\pi}{|a|}$$

Therefore:

$$f_1(x) = \cos^2(x) \Rightarrow T_1 = \frac{\pi}{1} = \pi$$

$$f_2(x) = -5\cos\left(\frac{2x}{3}\right) \Rightarrow T_2 = \frac{2\pi}{\frac{2}{3}} = 3\pi$$

The main period of the given expression is the least common multiple (LCM) of the main periods of the terms, as can be seen in the following:

$$T = \text{LCM}(\pi, 3\pi)$$

$$\Rightarrow T = 3\pi$$

Choice (3) is the answer.

4.10. From trigonometry, we know that:

$$f_1(x) = \sin^{2n}(ax), \forall n \in \mathbb{Z} \Rightarrow T_1 = \frac{\pi}{|a|}$$

$$f_2(x) = \cos^{2n+1}(ax), \forall n \in \mathbb{Z} \Rightarrow T_2 = \frac{2\pi}{|a|}$$

Therefore:

$$f_1(x) = \sin^4\left(\frac{3x}{5}\right) \Rightarrow T_1 = \frac{\pi}{\frac{3}{5}} = \frac{5\pi}{3}$$

$$f_2(x) = \cos^3\left(\frac{2x}{3}\right) \Rightarrow T_2 = \frac{2\pi}{\frac{2}{3}} = 3\pi$$

The main period of the given expression is: the least common multiple (LCM) of the main periods of the terms as follows:

$$T = \text{LCM}\left(\frac{5\pi}{3}, 3\pi\right)$$

$$\Rightarrow T = 15\pi$$

Choice (3) is the answer.

4.11. From trigonometry, we know that:

$$f_1(x) = \sin^{2n}(ax), \forall n \in \mathbb{Z} \Rightarrow T_1 = \frac{\pi}{|a|}$$

$$f_2(x) = \cos^{2n+1}(ax), \forall n \in \mathbb{Z} \Rightarrow T_2 = \frac{2\pi}{|a|}$$

Therefore:

$$f_1(x) = \sin^4\left(\frac{\pi x}{3}\right) \Rightarrow T_1 = \frac{\pi}{\frac{\pi}{3}} = 3$$

$$f_2(x) = \cos(\pi x) \Rightarrow T_2 = \frac{2\pi}{\pi} = 2$$

The main period of the given expression is: the least common multiple (LCM) of the main periods of the terms, as can be seen in the following:

$$T = \text{LCM}(3, 2)$$

$$\Rightarrow T = 6$$

Choice (4) is the answer.

4.12. From trigonometry, we know that:

$$y = \sin(kx) \Rightarrow T = \frac{2\pi}{|k|}$$

Therefore:

$$\Rightarrow \frac{3\pi}{4} = \frac{1}{2}\left(\frac{2\pi}{|k|}\right) \Rightarrow |k| = \frac{4}{3} \Rightarrow k = \pm\frac{4}{3}$$

Based on the graph and the function, the positive value of k is acceptable.

$$\Rightarrow k = \frac{4}{3}$$

Choice (4) is the answer (Fig. 4.1).

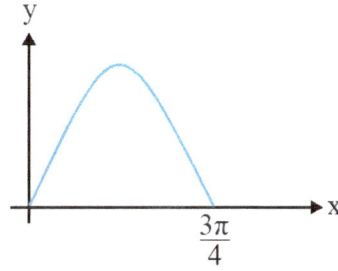

Figure 4.1 The graph of solution of problem 4.12

4.13. From trigonometry, we know that:

$$y = \cos\left(\pi ax + \frac{\pi}{2}\right) = -\sin(\pi ax)$$

$$y = \sin(mx) \Rightarrow T = \frac{2\pi}{|m|}$$

Therefore:

$$\Rightarrow 1 - \left(-\frac{1}{3}\right) = \frac{2\pi}{|\pi a|} \Rightarrow \frac{4}{3} = \frac{2}{|a|} \Rightarrow a = \pm\frac{3}{2}$$

Based on the graph and $y = -\sin(\pi ax)$, the positive value of a is acceptable.

$$\Rightarrow a = \frac{3}{2}$$

Choice (2) is the answer (Fig. 4.2).

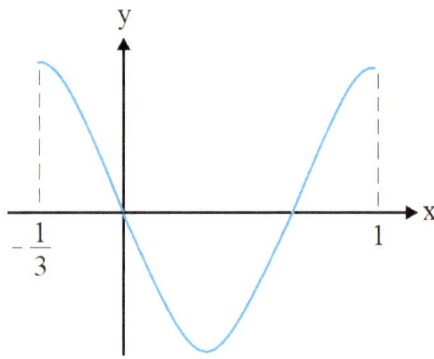

Figure 4.2 The graph of solution of problem 4.13

4.14. From trigonometry, we know that:

$$\sin(\alpha + 2n\pi) = \sin(\alpha), \forall n \in \mathbb{Z}$$

$$\cos(\alpha + 2n\pi) = \cos(\alpha), \forall n \in \mathbb{Z}$$

$$\tan(\alpha + n\pi) = \tan(\alpha), \forall n \in \mathbb{Z}$$

$$\cot(\alpha + n\pi) = \cot(\alpha), \forall n \in \mathbb{Z}$$

$$\sin(\pi - \alpha) = \sin(\alpha)$$

$$\cos(\pi - \alpha) = -\cos(\alpha)$$

$$\tan(-\alpha) = -\tan(\alpha)$$

$$\cot(-\alpha) = -\cot(\alpha)$$

Based on the information given in the problem, we have:

$$\alpha + \beta = 19\pi \Rightarrow \alpha = 19\pi - \beta$$

Therefore:

$$\sin(\alpha) = \sin(19\pi - \beta) = \sin(\pi - \beta) = \sin(\beta)$$

$$\cos(\alpha) = \cos(19\pi - \beta) = \cos(\pi - \beta) = -\cos(\beta)$$

$$\tan(\alpha) = \tan(19\pi - \beta) = \tan(-\beta) = -\tan(\beta)$$

$$\cot(\alpha) = \cot(19\pi - \beta) = \cot(-\beta) = -\cot(\beta)$$

Choice (1) is the answer.

4.15. From trigonometry, we know that:

$$\sin(\alpha + 2n\pi) = \sin(\alpha), \forall n \in \mathbb{Z}$$

$$\cos(\alpha + 2n\pi) = \cos(\alpha), \forall n \in \mathbb{Z}$$

$$\sin(\alpha + \pi) = -\sin(\alpha)$$

$$\sin(\alpha) + \sin(\beta) = 2\sin\left(\frac{\alpha + \beta}{2}\right)\cos\left(\frac{\alpha - \beta}{2}\right)$$

Therefore:

$$\sin(5\pi + x) + \sin\left(x - \frac{\pi}{3}\right) + \sin\left(x + \frac{7\pi}{3}\right) = \sin(x + \pi) + \sin\left(x - \frac{\pi}{3}\right) + \sin\left(x + \frac{\pi}{3}\right)$$

$$= -\sin(x) + 2\sin(x)\cos\left(\frac{\pi}{3}\right) = -\sin(x) + \sin(x)$$

$$\Rightarrow \sin(5\pi + x) + \sin\left(x - \frac{\pi}{3}\right) + \sin\left(x + \frac{7\pi}{3}\right) = 0$$

Choice (1) is the answer.

4.16. Let us assume:

$$\cos\left(\frac{2\pi}{3}\right) \triangleq -\frac{1}{2} \Rightarrow \text{arc}\left(\cos\left(-\frac{1}{2}\right)\right) = \frac{2\pi}{3}$$

$$\sin\left(\frac{-\pi}{3}\right) \triangleq -\frac{1}{2} \Rightarrow \text{arc}\left(\sin\left(\frac{-\sqrt{3}}{2}\right)\right) = \frac{-\pi}{3}$$

Therefore:

$$\sin\left[\text{arc}\left(\cos\left(-\frac{1}{2}\right)\right) + \text{arc}\left(\sin\left(\frac{-\sqrt{3}}{2}\right)\right)\right] = \sin\left[\frac{2\pi}{3} + \left(\frac{-\pi}{3}\right)\right] = \sin\frac{\pi}{3}$$

$$\Rightarrow \sin\left[\text{arc}\left(\cos\left(-\frac{1}{2}\right)\right) + \text{arc}\left(\sin\left(\frac{-\sqrt{3}}{2}\right)\right)\right] = \frac{\sqrt{3}}{2}$$

Choice (1) is the answer.

4.17. Let us assume:

$$\text{arc}\left(\tan\left(\frac{1}{2}\right)\right) \triangleq \alpha \Rightarrow \tan(\alpha) = \frac{1}{2}$$

We need to find the value of $2\arctan\left(\frac{1}{2}\right)$ which is equal to 2α.

From trigonometry, we know that:

$$\tan(2\alpha) = \frac{2\tan\alpha}{1 - \tan^2\alpha}$$

Hence:

$$\tan(2\alpha) = \frac{2 \times \frac{1}{2}}{1 - \left(\frac{1}{2}\right)^2} = \frac{4}{3}$$

$$\Rightarrow 2\alpha = \text{arc}\left(\tan\left(\frac{4}{3}\right)\right)$$

Choice (1) is the answer.

4.18. As we know:

$$e^{\ln a} = a$$

Moreover, from trigonometry of hyperbolic functions, we know that:

$$\cos\left(\frac{\pi}{3}\right) = \frac{1}{2}$$

$$\sinh x = \frac{e^x - e^{-x}}{2}$$

Thus, for $x = \ln 3$, we can write:

$$\sinh \ln 3 = \frac{e^{\ln 3} - e^{-\ln 3}}{2} = \frac{3 - \frac{1}{3}}{2} = \frac{4}{3}$$

Therefore,

$$\cos(\pi \sinh \ln 3) = \cos\left(\pi \times \frac{4}{3}\right) = \cos\left(\pi + \frac{\pi}{3}\right) = -\cos\left(\frac{\pi}{3}\right)$$

$$\Rightarrow \cos(\pi \sinh \ln 3) = -\frac{1}{2}$$

Choice (2) is the answer.

4.19. From trigonometry, we know that:

$$\sin(30^\circ) = \frac{1}{2}$$

$$\sin(\alpha) + \sin(\beta) = 2\sin\left(\frac{\alpha + \beta}{2}\right)\cos\left(\frac{\alpha - \beta}{2}\right)$$

Therefore:

$$\sin(50^\circ) + \sin(10^\circ) = m$$

$$\Rightarrow 2\sin(30^\circ)\cos(20^\circ) = m \Rightarrow 2 \times \frac{1}{2}\cos(20^\circ) = m \Rightarrow \cos(20^\circ) = m$$

Choice (2) is the answer.

4.20. From trigonometry, we know that:

$$\cos(\theta) = \frac{1 - \tan^2\left(\frac{\theta}{2}\right)}{1 + \tan^2\left(\frac{\theta}{2}\right)}$$

$$\tan(\theta) = \frac{\sin(\theta)}{\cos(\theta)}$$

$$\tan(45^\circ) = 1$$

Therefore:

$$\frac{\left(1 + \tan^2(5^\circ)\right)\sin(10^\circ)}{\left(1 - \tan^2(5^\circ)\right)\tan(10^\circ)} = \frac{1}{\cos(10^\circ)} \times \frac{\sin(10^\circ)}{\frac{\sin(10^\circ)}{\cos(10^\circ)}} = 1$$

$$\Rightarrow \frac{\left(1 + \tan^2(5^\circ)\right)\sin(10^\circ)}{\left(1 - \tan^2(5^\circ)\right)\tan(10^\circ)} = \tan(45^\circ)$$

Choice (4) is the answer.

4.21. From trigonometry, we know that:

$$\frac{1}{\cos^2(\alpha)} = 1 + \tan^2(\alpha)$$

$$\tan(\alpha) = \frac{1}{\cot(\alpha)}$$

Based on the information given in the problem, we have:

$$\cot(\alpha) = m$$

$$\cos(\alpha) = n$$

Therefore:

$$\frac{1}{\cos^2(\alpha)} = 1 + \frac{1}{\cot^2(\alpha)}$$

$$\Rightarrow \frac{1}{n^2} = 1 + \frac{1}{m^2} \xrightarrow{\times m^2 n^2} m^2 = m^2 n^2 + n^2$$

$$\Rightarrow m^2\left(1 - n^2\right) = n^2$$

Choice (2) is the answer.

4.22. From trigonometry, we can know that:

$$\sin^{2n+1}(ax), \forall n \in \mathbb{Z} \Rightarrow T = \frac{2\pi}{|a|}$$

$$\sin(\alpha)\cos(\beta) = \frac{1}{2}\left(\sin(\alpha + \beta) + \sin(\alpha - \beta)\right)$$

We need to change the product expression to the summation one, as follows:

$$y = \sin(3x)\cos(5x) + 11 \Rightarrow y = \frac{1}{2}\sin(8x) - \frac{1}{2}\sin(2x) + 11$$

Then:

$$\frac{1}{2}\sin(8x) \Rightarrow T_1 = \frac{2\pi}{8} = \frac{\pi}{4}$$

$$-\frac{1}{2}\sin(2x) \Rightarrow T_2 = \frac{2\pi}{2} = \pi$$

The main period of the given expression is: the least common multiple (LCM) of the main periods of the terms, as is presented in the following:

$$\Rightarrow T = \text{LCM}\left(\frac{\pi}{4}, \pi\right)$$

$$\Rightarrow T = \pi$$

Choice (1) is the answer.

4.23. From trigonometry, we know that:

$$\sin\left(\frac{4\pi}{3}\right) = \sin\left(\pi + \frac{\pi}{3}\right) = -\sin\left(\frac{\pi}{3}\right) = -\frac{\sqrt{3}}{2}$$

$$\text{arc}(\cos(-\alpha)) = \pi - \text{arc}(\cos(\alpha))$$

$$\text{arc}\left(\cos\left(\frac{\sqrt{3}}{2}\right)\right) = \frac{\pi}{6}$$

Therefore:

$$\arc\left(\cos\left(\sin\left(\frac{4\pi}{3}\right)\right)\right) = \arc\left(\cos\left(-\frac{\sqrt{3}}{2}\right)\right) = \pi - \arc\left(\cos\left(\frac{\sqrt{3}}{2}\right)\right) = \pi - \frac{\pi}{6}$$

$$\Rightarrow \arc\left(\cos\left(\sin\left(\frac{4\pi}{3}\right)\right)\right) = \frac{5\pi}{6}$$

Choice (2) is the answer.

4.24. From trigonometry, we know that:

$$\sin\left(\frac{17\pi}{5}\right) = \sin\left(4\pi - \frac{3\pi}{5}\right) = \sin\left(-\frac{3\pi}{5}\right)$$

$$\sin\left(-\frac{3\pi}{5}\right) \triangleq \alpha \Rightarrow \arc(\sin(\alpha)) = -\frac{3\pi}{5}$$

Therefore:

$$\arc\left(\sin\left(\sin\left(\frac{17\pi}{5}\right)\right)\right) = \arc\left(\sin\left(\sin\left(-\frac{3\pi}{5}\right)\right)\right) = \arc(\sin(\alpha))$$

$$\Rightarrow \arc\left(\sin\left(\sin\left(\frac{17\pi}{5}\right)\right)\right) = -\frac{3\pi}{5}$$

Choice (4) is the answer.

4.25. From trigonometry, we know that:

$$\cos\left(\frac{19\pi}{5}\right) = \cos\left(4\pi - \frac{\pi}{5}\right) = \cos\left(-\frac{\pi}{5}\right)$$

$$\cos\left(\frac{\pi}{5}\right) \triangleq \alpha \Rightarrow \arc(\cos(\alpha)) = \frac{\pi}{5}$$

Therefore:

$$\arc\left(\cos\left(\cos\left(\frac{19\pi}{5}\right)\right)\right) = \arc\left(\cos\left(\cos\left(-\frac{\pi}{5}\right)\right)\right) = \arc\left(\cos\left(\cos\left(\frac{\pi}{5}\right)\right)\right) = \arc(\cos(\alpha))$$

$$\Rightarrow \arc\left(\cos\left(\cos\left(\frac{19\pi}{5}\right)\right)\right) = \frac{\pi}{5}$$

Choice (1) is the answer.

4.26. From trigonometry, we know that:

$$\tan(2\alpha) = \frac{2\tan(\alpha)}{1 - \tan^2(\alpha)}$$

$$\arc\left(\tan\left(\frac{1}{2}\right)\right) \triangleq \alpha \Rightarrow \tan(\alpha) = \frac{1}{2}$$

Therefore:

$$\tan\left(2\mathrm{arc}\left(\tan\left(\frac{1}{2}\right)\right)\right) = \tan(2\alpha) = \frac{2\tan(\alpha)}{1 - \tan^2(\alpha)} = \frac{2 \times \frac{1}{2}}{1 - \left(\frac{1}{2}\right)^2} = \frac{1}{\frac{3}{4}}$$

$$\Rightarrow \tan\left(2\mathrm{arc}\left(\tan\left(\frac{1}{2}\right)\right)\right) = \frac{4}{3}$$

Choice (3) is the answer.

4.27. From trigonometry, we know that:

$$\sin^2(\alpha) + \cos^2(\alpha) = 1$$

$$1 + \tan^2(\alpha) = \frac{1}{\cos^2(\alpha)}$$

Therefore:

$$\mathrm{arc}\left(\sin\left(\frac{3}{5}\right)\right) \triangleq \alpha \Rightarrow \sin(\alpha) = \frac{3}{5} \Rightarrow \cos(\alpha) = \frac{4}{5}$$

$$\mathrm{arc}\left(\tan\left(\frac{3}{4}\right)\right) \triangleq \beta \Rightarrow \tan(\beta) = \frac{3}{4} \Rightarrow \cos(\beta) = \frac{4}{5} \Rightarrow \sin(\beta) = \frac{3}{5}$$

$$\Rightarrow \sin\left(\mathrm{arc}\left(\sin\left(\frac{3}{5}\right)\right) + \mathrm{arc}\left(\tan\left(\frac{3}{4}\right)\right)\right) = \sin(\alpha + \beta) = \sin(\alpha)\cos(\beta) + \cos(\alpha)\sin(\beta) = \frac{3}{5} \times \frac{4}{5} + \frac{4}{5} \times \frac{3}{5}$$

$$\Rightarrow \sin\left(\mathrm{arc}\left(\sin\left(\frac{3}{5}\right)\right) + \mathrm{arc}\left(\tan\left(\frac{3}{4}\right)\right)\right) = \frac{24}{25}$$

Choice (4) is the answer.

4.28. From trigonometry, we know that:

$$\mathrm{arc}(\cot(-\alpha)) = \pi - \mathrm{arc}(\cot(\alpha))$$

$$\mathrm{arc}(\cot(\alpha)) + \mathrm{arc}\left(\cot\left(\frac{1}{\alpha}\right)\right) = \frac{\pi}{2}$$

Therefore:

$$\mathrm{arc}\left(\cot\left(-\frac{4}{3}\right)\right) - \mathrm{arc}\left(\cot\left(\frac{3}{4}\right)\right) = \pi - \mathrm{arc}\left(\cot\left(\frac{4}{3}\right)\right) - \mathrm{arc}\left(\cot\left(\frac{3}{4}\right)\right) = \pi - \left(\mathrm{arc}\left(\cot\left(\frac{4}{3}\right)\right) + \mathrm{arc}\left(\cot\left(\frac{3}{4}\right)\right)\right) = \pi - \frac{\pi}{2}$$

$$\Rightarrow \mathrm{arc}\left(\cot\left(-\frac{4}{3}\right)\right) - \mathrm{arc}\left(\cot\left(\frac{3}{4}\right)\right) = \frac{\pi}{2}$$

Choice (3) is the answer.

4.29. From trigonometry, we know that:

$$\tan(\alpha + \beta) = \frac{\tan(\alpha) + \tan(\beta)}{1 - \tan(\alpha)\tan(\beta)}$$

$$\mathrm{arc}(\tan(5)) \triangleq \alpha \Rightarrow \tan(\alpha) = 5$$

$$\mathrm{arc}\left(\tan\left(\frac{3}{2}\right)\right) \triangleq \beta \Rightarrow \tan(\beta) = \frac{3}{2}$$

$$\mathrm{arc}(\tan(-1)) = \tan^{-1}(-1) = \frac{3\pi}{4}$$

Therefore:

$$\tan\left(\mathrm{arc}(\tan(5)) + \mathrm{arc}\left(\tan\left(\frac{3}{2}\right)\right)\right) = \tan(\alpha + \beta) = \frac{\tan(\alpha) + \tan(\beta)}{1 - \tan(\alpha)\tan(\beta)} = \frac{5 + \frac{3}{2}}{1 - \frac{15}{2}} = \frac{\frac{13}{2}}{-\frac{13}{2}} = -1$$

$$\Rightarrow \mathrm{arc}(\tan(5)) + \mathrm{arc}\left(\tan\left(\frac{3}{2}\right)\right) = \tan^{-1}(-1)$$

$$\Rightarrow \mathrm{arc}(\tan(5)) + \mathrm{arc}\left(\tan\left(\frac{3}{2}\right)\right) = \frac{3\pi}{4}$$

Choice (3) is the answer.

4.30. From trigonometry, we know that:

$$\sin^2(\alpha) + \cos^2(\alpha) = 1$$

$$\mathrm{arc}\left(\cos\left(\frac{3}{5}\right)\right) \triangleq \alpha \Rightarrow \cos(\alpha) = \frac{3}{5}$$

$$\mathrm{arc}\left(\sin\left(-\frac{4}{5}\right)\right) \triangleq \beta \Rightarrow \sin(\beta) = -\frac{4}{5}$$

Therefore:

$$\sin\left(\mathrm{arc}\left(\cos\left(\frac{3}{5}\right)\right)\right) + \cos\left(\mathrm{arc}\left(\sin\left(-\frac{4}{5}\right)\right)\right) = \sin(\alpha) + \cos(\beta) = \sqrt{1 - \left(\frac{3}{5}\right)^2} + \sqrt{1 - \left(-\frac{4}{5}\right)^2} = \frac{4}{5} + \frac{3}{5}$$

$$\Rightarrow \sin\left(\mathrm{arc}\left(\cos\left(\frac{3}{5}\right)\right)\right) + \cos\left(\mathrm{arc}\left(\sin\left(-\frac{4}{5}\right)\right)\right) = \frac{7}{5}$$

Choice (1) is the answer.

4.31. From trigonometry, we know that:

$$\cot(\theta) = \tan\left(\frac{\pi}{2} - \theta\right)$$

Based on the definition of $\tan(\theta)$ and $\cot(\theta)$, we can write:

$$\tan(\theta) = \frac{Opposite\ for\ \theta}{Adjacent\ for\ \theta} = \frac{HA}{OH} = \frac{HA}{1} = HA$$

$$\cot(\theta) = \tan\left(\widehat{HOB}\right) = \frac{Opposite\ for\ \widehat{HOB}}{Adjacent\ for\ \widehat{HOB}} = \frac{HB}{OH} = \frac{HB}{1} = HB$$

Choice (2) is the answer (Fig. 4.3).

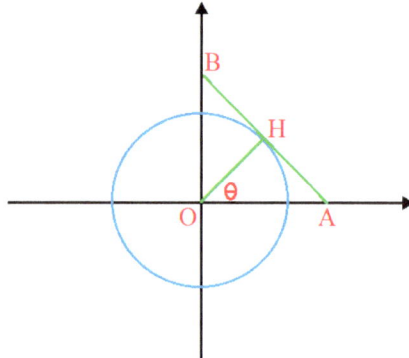

Figure 4.3 The graph of solution of problem 4.31

4.32. Based on the definition of $\sec(\theta)$ and $\cos(\theta)$, we can write:

$$\sec(\theta) = \frac{1}{\cos(\theta)} = \frac{1}{\frac{Adjacent\ for\ \theta}{Hypotenuse\ for\ \theta}} = \frac{1}{\frac{OA}{OB}} = \frac{1}{\frac{1}{OB}} = OB$$

Choice (3) is the answer (Fig. 4.4).

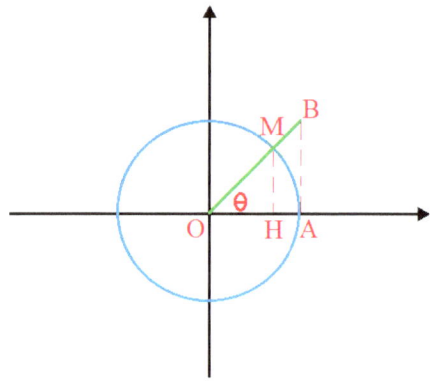

Figure 4.4 The graph of solution of problem 4.32

4.33. From trigonometry, we know that:

$$\sin(\theta) = \cos\left(\frac{\pi}{2} - \theta\right)$$

Based on the definition of $\csc(\theta)$ and $\sin(\theta)$, we can write:

$$\csc(\theta) = \frac{1}{\sin(\theta)} = \frac{1}{\cos\left(\widehat{COB}\right)} = \frac{1}{\frac{Adjacent\ for\ \widehat{COB}}{Hypotenuse\ for\ \widehat{COB}}} = \frac{1}{\frac{OC}{OB}} = \frac{1}{\frac{1}{OB}} = OB$$

Choice (2) is the answer (Fig. 4.5).

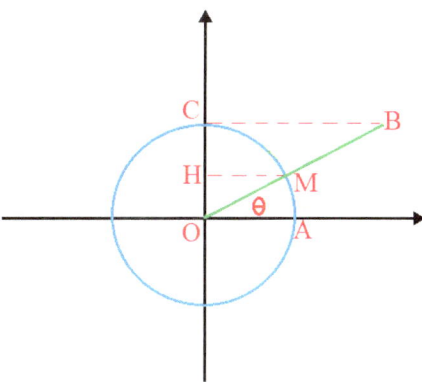

Figure 4.5 The graph of solution of problem 4.33

4.34. From trigonometry, we know that:

$$\tan(\alpha + \beta) = \frac{\tan(\alpha) + \tan(\beta)}{1 - \tan(\alpha)\tan(\beta)}$$

$$\text{arc}\left(\tan\left(\frac{2}{3}\right)\right) \triangleq \alpha \Rightarrow \tan(\alpha) = \frac{2}{3}$$

$$\text{arc}\left(\tan\left(\frac{1}{5}\right)\right) \triangleq \beta \Rightarrow \tan(\beta) = \frac{1}{5}$$

$$\text{arc}(\tan(1)) = \frac{\pi}{4}$$

Therefore:

$$\tan(\alpha + \beta) = \frac{\tan(\alpha) + \tan(\beta)}{1 - \tan(\alpha)\tan(\beta)} = \frac{\frac{2}{3} + \frac{1}{5}}{1 - \frac{2}{15}} = \frac{\frac{13}{15}}{\frac{13}{15}} = 1$$

$$\Rightarrow \alpha + \beta = \text{arc}(\tan(1))$$

$$\Rightarrow \alpha + \beta = \frac{\pi}{4}$$

Choice (2) is the answer.

4.35. From trigonometry, we know that:

$$\text{arc}(\tan(\alpha)) + \text{arc}\left(\tan\left(\frac{1}{\alpha}\right)\right) = \begin{cases} \dfrac{\pi}{2} & \text{if } \alpha > 0 \\[2mm] -\dfrac{\pi}{2} & \text{if } \alpha < 0 \end{cases}$$

$$\text{arc}(\cot(\alpha)) + \text{arc}(\cot(-\alpha)) = \pi$$

Therefore:

$$\text{arc}(\tan(m)) + \text{arc}\left(\tan\left(\frac{1}{m}\right)\right) + \text{arc}(\cot(m)) + \text{arc}(\cot(-m)) = \pi + \begin{cases} \dfrac{\pi}{2} & \text{if } m > 0 \\[2mm] -\dfrac{\pi}{2} & \text{if } m < 0 \end{cases}$$

$$\Rightarrow \mathrm{arc}(\tan(m)) + \mathrm{arc}\left(\tan\left(\frac{1}{m}\right)\right) + \mathrm{arc}(\cot(m)) + \mathrm{arc}(\cot(-m)) = \begin{cases} \dfrac{3\pi}{2} & \text{if } m > 0 \\ \dfrac{\pi}{2} & \text{if } m < 0 \end{cases}$$

Choice (2) is the answer.

4.36. From trigonometry, we know that:

$$\cos(\alpha - \beta) = \cos(\alpha)\cos(\beta) + \sin(\alpha)\sin(\beta)$$

Therefore:

$$-1 \le \cos(4x)\cos(2x) + \sin(4x)\sin(2x) \le 0$$

$$\Rightarrow -1 \le \cos(4x - 2x) \le 0 \Rightarrow -1 \le \cos(2x) \le 0$$

Since x is an acute angle:

$$\Rightarrow \frac{\pi}{2} \le 2x \le \pi \Rightarrow \frac{\pi}{4} \le x \le \frac{\pi}{2}$$

Choice (4) is the answer.

4.37. From trigonometry, we know that:

$$\tan(\alpha - \beta) = \frac{\tan(\alpha) - \tan(\beta)}{1 + \tan(\alpha)\tan(\beta)}$$

Therefore:

$$\tan(2y) = \tan((x+y) - (x-y)) = \frac{\tan(x+y) - \tan(x-y)}{1 + \tan(x+y)\tan(x-y)} = \frac{5 - 7}{1 + 5 \times 7} = \frac{-2}{1 + 35}$$

$$\Rightarrow \tan(2y) = \frac{-1}{18}$$

Choice (2) is the answer.

4.38. From trigonometry, we know that:

$$\sin(\alpha) + \cos(\alpha) = \sqrt{2}\sin\left(\alpha + \frac{\pi}{4}\right)$$

$$\sin(\alpha) - \cos(\alpha) = \sqrt{2}\sin\left(\alpha - \frac{\pi}{4}\right)$$

$$\sin\left(\frac{2\pi}{3}\right) = \frac{\sqrt{3}}{2}$$

$$\sin\left(\frac{\pi}{6}\right) = \frac{1}{2}$$

Therefore:

$$\frac{\sin\left(\frac{5\pi}{12}\right) + \cos\left(\frac{5\pi}{12}\right)}{\sin\left(\frac{5\pi}{12}\right) - \cos\left(\frac{5\pi}{12}\right)} = \frac{\sqrt{2}\sin\left(\frac{5\pi}{12} + \frac{\pi}{4}\right)}{\sqrt{2}\sin\left(\frac{5\pi}{12} - \frac{\pi}{4}\right)} = \frac{\sin\left(\frac{2\pi}{3}\right)}{\sin\left(\frac{\pi}{6}\right)} = \frac{\frac{\sqrt{3}}{2}}{\frac{1}{2}}$$

$$\Rightarrow \frac{\sin\left(\frac{5\pi}{12}\right) + \cos\left(\frac{5\pi}{12}\right)}{\sin\left(\frac{5\pi}{12}\right) - \cos\left(\frac{5\pi}{12}\right)} = \sqrt{3}$$

Choice (1) is the answer.

4.39. From trigonometry, we know that:

$$y = a\sin(mx) \Rightarrow T = \frac{2\pi}{|m|}$$

Therefore:

$$\Rightarrow 6 = \frac{2\pi}{|b\pi|} \Rightarrow |b| = \frac{1}{3} \Rightarrow b = \pm\frac{1}{3}$$

Based on the graph and the function, the positive value of b is acceptable.

$$\Rightarrow b = \frac{1}{3}$$

Moreover, based on $y = a\sin(b\pi x)$ and the given graph, it is concluded that $a = 2$. Therefore:

$$\Rightarrow a + b = 2 + \frac{1}{3}$$

$$\Rightarrow a + b = \frac{7}{3}$$

Choice (3) is the answer (Fig. 4.6).

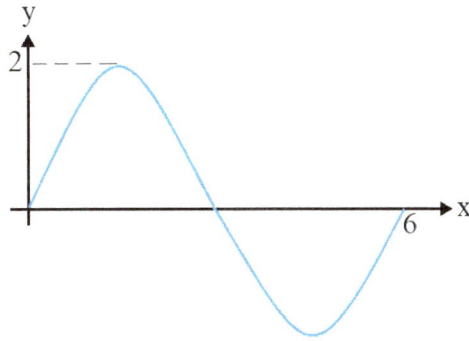

Figure 4.6 The graph of solution of problem 4.39

4.40. From trigonometry, we know that:

$$y = a\sin(mx) \Rightarrow T = \frac{2\pi}{|m|}$$

Therefore:

$$\Rightarrow 3 = 3 \times \frac{2\pi}{|b\pi|} = 1 \Rightarrow |b| = 2 \Rightarrow b = \pm 2$$

Based on the graph and the given function, the negative value of b is accepted.

$$\Rightarrow b = -2$$

In addition, based on $y = a\sin(b\pi x)$ and the given graph, it is clear that $a = 3$. Therefore:

$$\Rightarrow a \times b = 3 \times (-2)$$

$$\Rightarrow a \times b = -6$$

Choice (1) is the answer (Fig. 4.7).

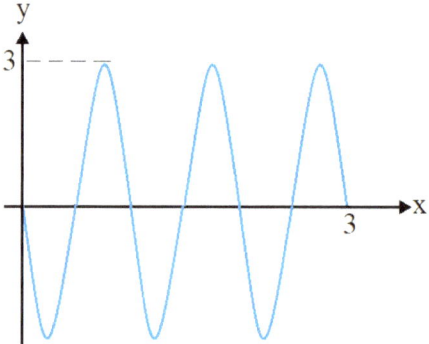

Figure 4.7 The graph of solution of problem 4.40

4.41. From trigonometry, we know that:

$$y = a\sin\left(\frac{\pi}{2} + b\pi x\right) = a\cos(b\pi x)$$

$$y = a\cos(mx) \Rightarrow T = \frac{2\pi}{|m|}$$

Therefore:

$$\Rightarrow 3.5 - (-2.5) = 3 \times \frac{2\pi}{|b\pi|} \Rightarrow 6 = \frac{6}{|b|} \Rightarrow |b| = 1 \Rightarrow b = \pm 1$$

Based on the graph and $y = a\cos(b\pi x)$, the positive value of b is accepted.

$$\Rightarrow b = 1$$

In addition, based on $y = a \cos(b\pi x)$ and the given graph, it is clear that $a = 2$. Therefore:

$$\Rightarrow a \times b = 2 \times 1$$

$$\Rightarrow a \times b = 2$$

Choice (1) is the answer (Fig. 4.8).

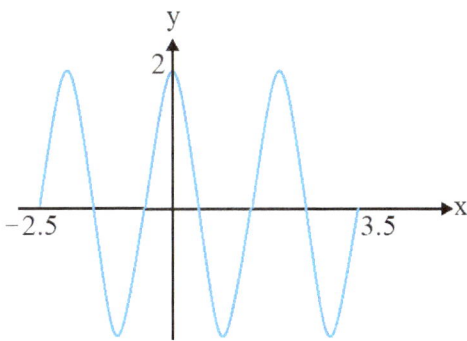

Figure 4.8 The graph of solution of problem 4.41

4.42. From trigonometry, we know that:

$$\cos(\alpha) = \sin\left(\frac{\pi}{2} - \alpha\right)$$

$$\sin(\alpha) = \cos\left(\frac{\pi}{2} - \alpha\right)$$

$$\sin(2\alpha) = 2\sin(\alpha)\cos(\alpha)$$

Therefore:

$$\frac{\cos(5°)\cos(10°)\cos(20°)}{\cos(50°)} = \frac{\cos(5°)\cos(10°)\cos(20°)}{\sin(40°)} = \frac{\cos(5°)\cos(10°)\cos(20°)}{2\sin(20°)\cos(20°)}$$

$$= \frac{\cos(5°)\cos(10°)}{2 \times 2\sin(10°)\cos(10°)} = \frac{\cos(5°)}{4 \times 2\sin(5°)\cos(5°)} = \frac{1}{8\sin(5°)}$$

$$\Rightarrow \frac{\cos(5°)\cos(10°)\cos(20°)}{\cos(50°)} = \frac{1}{8\cos(85°)}$$

Choice (2) is the answer.

4.43. From trigonometry, we know that:

$$\sin(x) = \frac{2\tan\left(\frac{x}{2}\right)}{1 + \tan^2\left(\frac{x}{2}\right)}$$

$$\cos(x) = \frac{1 - \tan^2\left(\frac{x}{2}\right)}{1 + \tan^2\left(\frac{x}{2}\right)}$$

Therefore:

$$\sin(x) + \cos(x) = \frac{7}{5} \Rightarrow \frac{2\tan\left(\frac{x}{2}\right)}{1 + \tan^2\left(\frac{x}{2}\right)} + \frac{1 - \tan^2\left(\frac{x}{2}\right)}{1 + \tan^2\left(\frac{x}{2}\right)} = \frac{7}{5} \Rightarrow \frac{2\tan\left(\frac{x}{2}\right) + 1 - \tan^2\left(\frac{x}{2}\right)}{1 + \tan^2\left(\frac{x}{2}\right)} = \frac{7}{5}$$

$$\Rightarrow 12\,\tan^2\left(\frac{x}{2}\right) - 10\tan\left(\frac{x}{2}\right) + 2 = 0 \Rightarrow \tan\left(\frac{x}{2}\right) = \frac{10 \pm \sqrt{10^2 - 4 \times 12 \times 2}}{24} = \frac{10 \pm 2}{24}$$

$$\Rightarrow \tan\left(\frac{x}{2}\right) = \frac{1}{2} \; or \; \frac{1}{3}$$

Choice (2) is the answer.

4.44. From trigonometry, we know that:

$$\sin(2\alpha) = 2\sin(\alpha)\cos(\alpha)$$

$$\sin^2(\alpha) + \cos^2(\alpha) = 1$$

$$\cos^2(\alpha) - \sin^2(\alpha) = \cos(2\alpha)$$

$$\cot(2\alpha) = \frac{\cos(2\alpha)}{\sin(2\alpha)}$$

In addition, from factoring rule, we know that:

$$a^4 - b^4 = \left(a^2 - b^2\right)\left(a^2 + b^2\right)$$

Therefore:

$$\frac{\sin^4(\alpha) - \cos^4(\alpha)}{\sin(\alpha)\cos(\alpha)} = \frac{\left(\sin^2(\alpha) - \cos^2(\alpha)\right)\left(\sin^2(\alpha) + \cos^2(\alpha)\right)}{\sin(\alpha)\cos(\alpha)} = \frac{-\cos(2\alpha) \times 1}{\frac{1}{2}\sin(2\alpha)}$$

$$\Rightarrow \frac{\sin^4(\alpha) - \cos^4(\alpha)}{\sin(\alpha)\cos(\alpha)} = -2\cot(2\alpha)$$

Choice (2) is the answer.

4.45. Based on the information given in the problem, we have:

$$\sin^4(\alpha) + \cos^4(\alpha) = \frac{1}{2}$$

From trigonometry, we know that:

$$\sin(2\alpha) = 2\sin(\alpha)\cos(\alpha)$$

$$\sin^2(\alpha) + \sin^2(\alpha) = 1 \Rightarrow \left(\sin^2(\alpha) + \sin^2(\alpha)\right)^2 = 1$$

$$\Rightarrow \sin^4(\alpha) + \cos^4(\alpha) + 2\sin^2(\alpha)\cos^2(\alpha) = 1$$

Therefore:

$$\frac{1}{2} + 2\sin^2(\alpha)\cos^2(\alpha) = 1 \Rightarrow 4\sin^2(\alpha)\cos^2(\alpha) = 1 \Rightarrow (2\sin(\alpha)\cos(\alpha))^2 = 1$$

$$\Rightarrow \sin^2(2\alpha) = \pm 1 \Rightarrow \cos^2(2\alpha) = 0 \Rightarrow \frac{\cos^2(2\alpha)}{\sin^2(2\alpha)} = 0$$

$$\Rightarrow \cot^2(2\alpha) = 0$$

Choice (3) is the answer.

4.46. From trigonometry, we know that:

$$\sin(2x) = 2\sin(x)\cos(x)$$

$$\cos(2x) = \cos^2(x) - \sin^2(x)$$

Therefore:

$$\sin^3(x)\cos(x) - \cos^3(x)\sin(x) + 3\sin^2(x)\cos^2(x)$$

$$= \sin(x)\cos(x)\left(\sin^2(x) - \cos^2(x)\right) + \frac{3}{4} \times 4\sin^2(x)\cos^2(x)$$

$$= \frac{1}{2}\sin(2x)(-\cos(2x)) + \frac{3}{4}\sin^2(2x)$$

$$= -\frac{1}{4}\sin(4x) + \frac{3}{4}\sin^2(2x)$$

For $x = \dfrac{3\pi}{8}$, we have:

$$-\frac{1}{4}\sin\left(4 \times \frac{3\pi}{8}\right) + \frac{3}{4}\sin^2\left(2 \times \frac{3\pi}{8}\right) = -\frac{1}{4}(-1) + \frac{3}{4}\left(\frac{\sqrt{2}}{2}\right)^2 = \frac{1}{4} + \frac{3}{8} = \frac{5}{8}$$

Choice (2) is the answer.

4.47. From trigonometry, we know that:

$$\sin(\alpha) + \sin(\beta) = 2\sin\left(\frac{\alpha + \beta}{2}\right)\cos\left(\frac{\alpha - \beta}{2}\right)$$

$$\cos(\alpha) + \cos(\beta) = 2\cos\left(\frac{\alpha + \beta}{2}\right)\cos\left(\frac{\alpha - \beta}{2}\right)$$

Therefore:

$$\frac{\sin(2\alpha) + \sin(5\alpha) + \sin(8\alpha)}{\cos(2\alpha) + \cos(5\alpha) + \cos(8\alpha)} = \frac{\sin(8\alpha) + \sin(2\alpha) + \sin(5\alpha)}{\cos(8\alpha) + \cos(2\alpha) + \cos(5\alpha)}$$

$$= \frac{2\sin(5\alpha)\cos(3\alpha) + \sin(5\alpha)}{2\cos(5\alpha)\cos(3\alpha) + \cos(5\alpha)} = \frac{\sin(5\alpha)(2\cos(3\alpha) + 1)}{\cos(5\alpha)(2\cos(3\alpha) + 1)} = \tan(5\alpha)$$

$$\xrightarrow{\alpha = \frac{\pi}{15}} \tan\left(5 \times \frac{\pi}{15}\right) = \tan\left(\frac{\pi}{3}\right) = \sqrt{3}$$

Choice (3) is the answer.

4.48. From trigonometry, we know that:

$$\sin^2(x) + \cos^2(x) = 1$$

$$2\sin(x)\cos(x) = \sin(2x)$$

In addition, from factoring rule, we know that:

$$(a+b)(a-b) = a^2 - b^2$$

Therefore:

$$(\sin(x) - \cos(x) + 2)(\sin(x) - \cos(x) - 2)$$

$$= (\sin(x) - \cos(x))^2 - 4 = \sin^2(x) + \cos^2(x)$$

$$- 2\sin(x)\cos(x) - 4 = 1 - \sin(2x) - 4$$

$$= -3 - \sin(2x)$$

$$\xrightarrow{x = \frac{\pi}{12}} - 3 - \sin\left(\frac{\pi}{6}\right) = -3 - \frac{1}{2} = -\frac{7}{2}$$

Choice (4) is the answer.

4.49. From trigonometry, we know that:

$$\tan(\alpha)\cot(\alpha) = 1$$

$$1 + \tan^2(\alpha) = \frac{1}{\cos^2(\alpha)}$$

$$1 + \cot^2(\alpha) = \frac{1}{\sin^2(\alpha)}$$

$$\sin^2(\alpha) + \cos^2(\alpha) = 1$$

Therefore:

$$4\sin^2(\alpha)\cos^2(\alpha)(\tan(\alpha) + \cot(\alpha))^2$$

$$= 4\sin^2(\alpha)\cos^2(\alpha)\left(\tan^2(\alpha) + \cot^2(\alpha) + 2\tan(\alpha)\cot(\alpha)\right)$$

$$= 4\sin^2(\alpha)\cos^2(\alpha)\left(1 + \tan^2(\alpha) + 1 + \cot^2(\alpha)\right)$$

$$= 4\sin^2(\alpha)\cos^2(\alpha)\left(\frac{1}{\cos^2(\alpha)}+\frac{1}{\sin^2(\alpha)}\right)$$

$$= 4\sin^2(\alpha)\cos^2(\alpha)\left(\frac{\sin^2(\alpha)+\cos^2(\alpha)}{\sin^2(\alpha)\cos^2(\alpha)}\right)$$

$$= 4\sin^2(\alpha)\cos^2(\alpha)\left(\frac{1}{\sin^2(\alpha)\cos^2(\alpha)}\right) = 4$$

Choice (4) is the answer.

4.50. From trigonometry, we know the common solution of the equations below.

$$\sin(\alpha) = 0 \Rightarrow \alpha = k\pi, \forall k \in \mathbb{Z}$$

$$\cos(\alpha) = 0 \Rightarrow \alpha = k\pi + \frac{\pi}{2}, \forall k \in \mathbb{Z}$$

$$\tan(\alpha) = -1 \Rightarrow \alpha = k\pi - \frac{\pi}{4}, \forall k \in \mathbb{Z}$$

Hence:

$$\sin(\pi x)\cos^2(\pi x) + \sin^2(\pi x)\cos(\pi x) = 0 \Rightarrow \sin(\pi x)\cos(\pi x)(\cos(\pi x) + \sin(\pi x)) = 0$$

$$\Rightarrow \begin{cases} \sin(\pi x) = 0 \Rightarrow \pi x = k\pi \Rightarrow x = k \xrightarrow{-2 \le x \le 2} x = -2, -1, 0, 1, 2 \\ \cos(\pi x) = 0 \Rightarrow \pi x = k\pi + \frac{\pi}{2} \Rightarrow x = k + \frac{1}{2} \xrightarrow{-2 \le x \le 2} x = -\frac{3}{2}, -\frac{1}{2}, \frac{1}{2}, \frac{3}{2} \\ \sin(\pi x) + \cos(\pi x) = 0 \Rightarrow \tan(\pi x) = -1 \Rightarrow \pi x = k\pi - \frac{\pi}{4} \Rightarrow x = k - \frac{1}{4} \xrightarrow{-2 \le x \le 2} x = -\frac{5}{4}, -\frac{1}{4}, \frac{3}{4}, \frac{7}{4} \end{cases}$$

Therefore, the number of roots of the equation is: $5 + 4 + 4 = 13$.

Choice (3) is the answer.

4.51. From trigonometry, we know that:

$$y = a + b\cos(mx) \Rightarrow T = \frac{2\pi}{|m|}$$

Therefore:

$$\Rightarrow 4\pi = \frac{2\pi}{|m|} \Rightarrow |m| = \frac{1}{2} \Rightarrow m = \pm\frac{1}{2}$$

Based on the graph and the function, the positive value of m is accepted.

$$\Rightarrow m = \frac{1}{2} \Rightarrow y = \frac{1}{2} + 2\cos\left(\frac{1}{2}x\right)$$

The value of function for $x = \frac{16\pi}{3}$ is:

$$y\left(\frac{16\pi}{3}\right) = \frac{1}{2} + 2\cos\left(\frac{1}{2}\times\frac{16\pi}{3}\right) = \frac{1}{2} + 2\cos\left(\frac{8\pi}{3}\right) = \frac{1}{2} + 2\cos\left(2\pi + \frac{2\pi}{3}\right) = \frac{1}{2} + 2\cos\left(\frac{2\pi}{3}\right)$$

$$= \frac{1}{2} + 2\cos\left(\pi - \frac{\pi}{3}\right) = \frac{1}{2} - 2\cos\left(\frac{\pi}{3}\right) = \frac{1}{2} - 2\left(\frac{1}{2}\right) = -\frac{1}{2}$$

Choice (1) is the answer (Fig. 4.9).

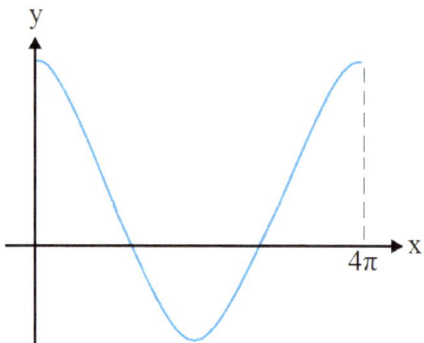

Figure 4.9 The graph of solution of problem 4.51

4.52. From trigonometry, we know that:

$$y = a + b\sin(mx) \Rightarrow T = \frac{2\pi}{|m|}$$

Therefore:

$$\Rightarrow \frac{2\pi}{3} = \frac{2\pi}{|m|} \Rightarrow |m| = 3 \Rightarrow m = \pm 3$$

Based on the graph and the given function, the positive value of m is accepted.

$$\Rightarrow m = 3 \Rightarrow y = 1 - \sin(3x)$$

The value of function for $x = \frac{7\pi}{6}$ is:

$$y\left(\frac{7\pi}{6}\right) = 1 - \sin\left(3 \times \frac{7\pi}{6}\right) = 1 - \sin\left(\frac{7\pi}{2}\right) = 1 - \sin\left(2\pi + \frac{3\pi}{2}\right) = 1 - \sin\left(\frac{3\pi}{2}\right) = 1 - (-1) = 2$$

Choice (4) is the answer (Fig. 4.10).

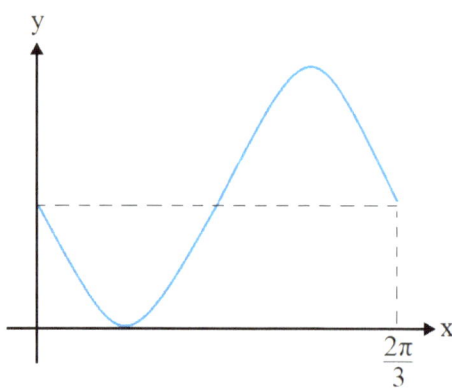

Figure 4.10 The graph of solution of problem 4.52

4.53. From trigonometry, we know that:

$$y = a + b\sin(mx) \Rightarrow T = \frac{2\pi}{|m|}$$

Therefore:

$$\Rightarrow 5 - 1 = \frac{2\pi}{|b\pi|} \Rightarrow b = \pm\frac{1}{2}$$

Based on the graph and the given function, the positive value of m is accepted.

$$\Rightarrow b = -\frac{1}{2} \Rightarrow y = a + \sin\left(-\frac{\pi}{2}x\right)$$

By testing the point of (0, 3) in the function, we have:

$$3 = a + \sin\left(-\frac{\pi}{2} \times 0\right) \Rightarrow a = 3 \Rightarrow y = 3 + \sin\left(-\frac{\pi}{2}x\right)$$

The value of function for $x = \frac{25}{3}$ is:

$$y\left(\frac{25}{3}\right) = 3 + \sin\left(-\frac{\pi}{2} \times \frac{25}{3}\right) = 3 + \sin\left(-4\pi - \frac{\pi}{6}\right)$$

$$= 3 + \sin\left(-\frac{\pi}{6}\right) = 3 - \frac{1}{2} = 2.5$$

Choice (2) is the answer (Fig. 4.11).

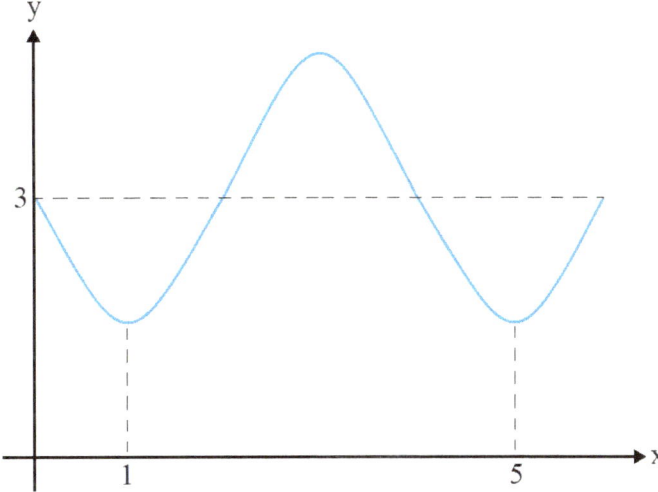

Figure 4.11   The graph of solution of problem 4.53

4.54. By testing the point of (0, 0) in the function, we have:

$$0 = a + b\cos\left(\frac{\pi}{2} \times 0\right) \Rightarrow a + b = 0 \tag{1}$$

Based on the function and the graph given in the problem, we can write:

$$y_{max} = a + |b| \Rightarrow a + |b| = 4 \tag{2}$$

The assumption of $b < 0$ is not acceptable because it results in the equations with an impossible solution, as can be seen in the following:

$$\xrightarrow{Using\ (1),(2)} \begin{cases} a + b = 0 \\ a + b = 4 \end{cases} \Rightarrow \text{Impossible}$$

However, for the assumption of $b > 0$, we have:

$$\xrightarrow{Using\ (1),(2)} \begin{cases} a + b = 0 \\ a - b = 4 \end{cases} \Rightarrow 2b = -4 \Rightarrow b = -2$$

Choice (1) is the answer (Fig. 4.12).

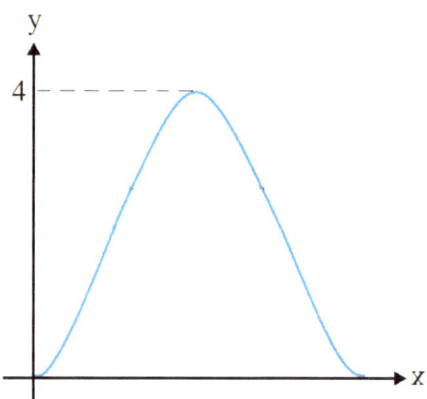

Figure 4.12 The graph of solution of problem 4.54

4.55. From trigonometry, we know that:

$$y = 1 + a\sin(mx) \Rightarrow T = \frac{2\pi}{|m|}$$

Therefore:

$$\Rightarrow \frac{4}{3} = 2 \times \frac{2\pi}{|b\pi|} \Rightarrow |b| = 3 \Rightarrow b = \pm 3$$

Based on the function and the graph given in the problem, we can write:

$$y_{min} = 1 - |a| \Rightarrow -1 = 1 - |a| \Rightarrow |a| = 2 \Rightarrow a = \pm 2$$

Based on the graph and the given function, both of a and b must be either positive or negative. Hence:

$$\begin{cases} a=2 \\ b=3 \end{cases} \Rightarrow a+b=5$$

$$\begin{cases} a=-2 \\ b=-3 \end{cases} \Rightarrow a+b=-5$$

Only $a+b=5$ exists in the choices. Choice (3) is the answer (Fig. 4.13).

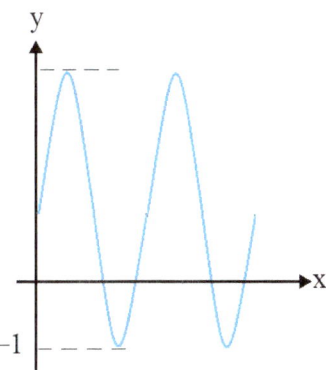

Figure 4.13 The graph of solution of problem 4.55

4.56. From trigonometry, we know that:

$$\cos\left(\alpha + \frac{\pi}{2}\right) = -\sin(\alpha)$$

Therefore:

$$y = a - 2\cos\left(bx + \frac{\pi}{2}\right) = a + 2\sin(bx)$$

In addition, from trigonometry, we know that:

$$y = a + 2\sin(bx) \Rightarrow T = \frac{2\pi}{|b|} \Rightarrow \frac{13\pi}{18} - \frac{\pi}{18} = \frac{2\pi}{|b|} \Rightarrow |b| = 3 \Rightarrow b = \pm 3$$

Based on the graph and the simplified function, i.e., $y = a + 2\sin(bx)$, the positive value of b is acceptable.

$$\Rightarrow b = 3$$

Based on the simplified function and the given graph, we can write:

$$y_{max} = a + 2 \Rightarrow 1 = a + 2 \Rightarrow a = -1$$

Hence:

$$a + b = -1 + 3 = 2$$

Choice (4) is the answer (Fig. 4.14).

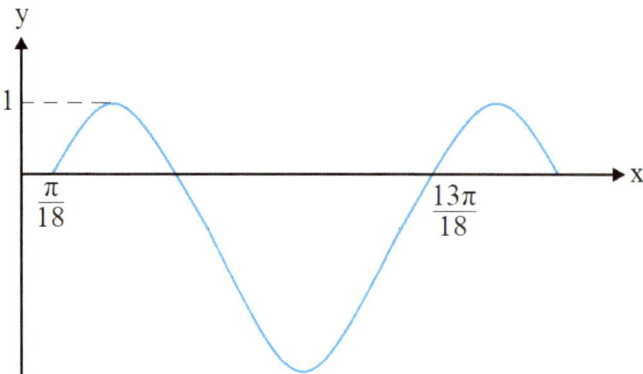

Figure 4.14 The graph of solution of problem 4.56

4.57. From trigonometry, we know that:

$$\tan\left(45^{\circ}\right) = 1$$

$$\tan(\alpha - \beta) = \frac{\tan(\alpha) - \tan(\beta)}{1 + \tan(\alpha)\tan(\beta)}$$

$$\cot(\alpha) = \frac{1}{\tan(\alpha)}$$

In addition, based on the information given in the problem, we have:

$$\tan\left(\alpha + 20^{\circ}\right) = \frac{3}{4}$$

Therefore:

$$\cot\left(25^{\circ} - \alpha\right) = \frac{1}{\tan\left(25^{\circ} - \alpha\right)} = \frac{1}{\tan\left(45^{\circ} - \left(\alpha + 20^{\circ}\right)\right)}$$

$$= \frac{1 + \tan\left(45^{\circ}\right)\tan\left(\alpha + 20^{\circ}\right)}{\tan\left(45^{\circ}\right) - \tan\left(\alpha + 20^{\circ}\right)} = \frac{1 + \tan\left(\alpha + 20^{\circ}\right)}{1 - \tan\left(\alpha + 20^{\circ}\right)} = \frac{1 + \frac{3}{4}}{1 - \frac{3}{4}}$$

$$\Rightarrow \cot\left(25^{\circ} - \alpha\right) = 7$$

Choice (3) is the answer.

4.58. From trigonometry, we know that:

$$\cos(\alpha) = -\sqrt{1 - \sin^{2}(\alpha)} \text{ for an obtuse angle}$$

$$\tan(\alpha) = \frac{\sin(\alpha)}{\cos(\alpha)}$$

$$\tan\left(\frac{\pi}{4}\right) = 1$$

$$\tan(\alpha + \beta) = \frac{\tan(\alpha) + \tan(\beta)}{1 - \tan(\alpha)\tan(\beta)}$$

In addition, based on the information given in the problem, we have:

$$\sin(\alpha) = \frac{3}{5}$$

Therefore:

$$\cos(\alpha) = -\sqrt{1 - \sin^2(\alpha)} = -\sqrt{1 - \left(\frac{3}{5}\right)^2} = -\frac{4}{5}$$

$$\tan(\alpha) = \frac{\sin(\alpha)}{\cos(\alpha)} = \frac{\frac{3}{5}}{-\frac{4}{5}} = -\frac{3}{4}$$

$$\tan\left(\frac{\pi}{4} + \alpha\right) = \frac{\tan\left(\frac{\pi}{4}\right) + \tan(\alpha)}{1 - \tan\left(\frac{\pi}{4}\right)\tan(\alpha)} = \frac{1 + \tan(\alpha)}{1 - \tan(\alpha)} = \frac{1 + \left(-\frac{3}{4}\right)}{1 - \left(-\frac{3}{4}\right)} = \frac{\frac{1}{4}}{\frac{7}{4}}$$

$$\Rightarrow \tan\left(\frac{\pi}{4} + \alpha\right) = \frac{1}{7}$$

Choice (3) is the answer.

4.59. From trigonometry, we know that:

$$\tan\left(\frac{\pi}{2} - \alpha\right) = \cot(\alpha)$$

$$\cot(\alpha) = \frac{1}{\tan(\alpha)}$$

$$\tan\left(\frac{\pi}{4}\right) = 1$$

$$\tan(\alpha - \beta) = \frac{\tan(\alpha) - \tan(\beta)}{1 + \tan(\alpha)\tan(\beta)}$$

In addition, based on the information given in the problem, we have:

$$\tan\left(\frac{\pi}{2} - \alpha\right) = \frac{2}{3}$$

Therefore:

$$\frac{2}{3} = \tan\left(\frac{\pi}{2} - \alpha\right) = \cot(\alpha) = \frac{1}{\tan(\alpha)} \Rightarrow \tan(\alpha) = \frac{3}{2}$$

$$\tan\left(\frac{\pi}{4} - \alpha\right) = \frac{\tan\left(\frac{\pi}{4}\right) - \tan(a)}{1 + \tan\left(\frac{\pi}{4}\right)\tan(a)} = \frac{1 - \tan(\alpha)}{1 + \tan(\alpha)} = \frac{1 - \frac{3}{2}}{1 + \frac{3}{2}} = \frac{-\frac{1}{2}}{\frac{5}{2}}$$

$$\Rightarrow \tan\left(\frac{\pi}{4} - \alpha\right) = -\frac{1}{5}$$

Choice (2) is the answer.

4.60. From trigonometry, we know that:

$$\tan(\alpha + \beta) = \frac{\tan(\alpha) + \tan(\beta)}{1 - \tan(\alpha)\tan(\beta)}$$

In addition, based on the information given in the problem, we have:

$$\tan(a + b) = \frac{2}{5}$$

$$\tan(a - b) = \frac{3}{7}$$

Therefore:

$$\tan(2a) = \tan((a + b) + (a - b)) = \frac{\tan(a + b) + \tan(a - b)}{1 - \tan(a + b)\tan(a - b)} = \frac{\frac{2}{5} + \frac{3}{7}}{1 - \frac{2}{5} \times \frac{3}{7}} = \frac{\frac{29}{35}}{\frac{29}{35}}$$

$$\Rightarrow \tan(2a) = 1$$

Choice (4) is the answer.

4.61. From trigonometry, we know that:

$$\sin(\alpha) = \cos\left(\frac{\pi}{2} - \alpha\right)$$

$$\cos(\alpha) = \cos(-\alpha)$$

$$\tan\left(\frac{\pi}{4}\right) = 1$$

$$\tan(\alpha - \beta) = \frac{\tan(\alpha) - \tan(\beta)}{1 + \tan(\alpha)\tan(\beta)}$$

Therefore:

$$\sin\left(\frac{\pi}{4} + x\right) = \cos\left(\frac{\pi}{2} - \left(\frac{\pi}{4} - x\right)\right) = \cos\left(\frac{\pi}{4} - x\right) = \cos\left(x - \frac{\pi}{4}\right)$$

$$\Rightarrow 2 = \frac{\sin\left(x - \frac{\pi}{4}\right)}{\sin\left(x + \frac{\pi}{4}\right)} = \frac{\sin\left(x - \frac{\pi}{4}\right)}{\cos\left(x - \frac{\pi}{4}\right)}$$

$$\Rightarrow \tan\left(x - \frac{\pi}{4}\right) = 2 \Rightarrow \frac{\tan(x) - \tan\left(\frac{\pi}{4}\right)}{1 + \tan(x)\tan\left(\frac{\pi}{4}\right)} = 2$$

$$\Rightarrow \frac{\tan(x) - 1}{1 + \tan(x)} = 2 \Rightarrow \tan(x) - 1 = 2 + 2\tan(x)$$

$$\Rightarrow \tan(x) = -3$$

Choice (1) is the answer.

4.62. From trigonometry, we know that:

$$\tan\left(\frac{\pi}{4}\right) = 1$$

$$\tan(\alpha + \beta) = \frac{\tan(\alpha) + \tan(\beta)}{1 - \tan(\alpha)\tan(\beta)}$$

Moreover, based on the information given in the problem, we have:

$$\alpha + \beta = \frac{\pi}{4}$$

If we calculate the tangent value of each side of the abovementioned relation, we will have:

$$\tan(\alpha + \beta) = \tan\left(\frac{\pi}{4}\right) \Rightarrow \frac{\tan(\alpha) + \tan(\beta)}{1 - \tan(\alpha)\tan(\beta)} = 1 \Rightarrow \tan(\alpha) + \tan(\beta) = 1 - \tan(\alpha)\tan(\beta)$$

Therefore:

$$(1 + \tan(\alpha))(1 + \tan(\beta)) = 1 + \tan(\alpha) + \tan(\beta) + \tan(\alpha)\tan(\beta) = 1 + (1 - \tan(\alpha)\tan(\beta)) + \tan(\alpha)\tan(\beta)$$

$$\Rightarrow (1 + \tan(\alpha))(1 + \tan(\beta)) = 2$$

Choice (2) is the answer.

4.63. From trigonometry, we know that:

$$\tan\left(\frac{\pi}{4}\right) = 1$$

$$\tan(\alpha + \beta) = \frac{\tan(\alpha) + \tan(\beta)}{1 - \tan(\alpha)\tan(\beta)}$$

$$\tan(\alpha - \beta) = \frac{\tan(\alpha) - \tan(\beta)}{1 + \tan(\alpha)\tan(\beta)}$$

$$\tan(2\alpha) = \frac{2\tan(\alpha)}{1 - \tan^2(\alpha)}$$

Therefore:

$$\tan\left(\frac{\pi}{4}+\alpha\right)-\tan\left(\frac{\pi}{4}-\alpha\right)=\frac{\tan\left(\frac{\pi}{4}\right)+\tan(\alpha)}{1-\tan\left(\frac{\pi}{4}\right)\tan(\alpha)}-\frac{\tan\left(\frac{\pi}{4}\right)-\tan(\alpha)}{1+\tan\left(\frac{\pi}{4}\right)\tan(\alpha)}$$

$$=\frac{1+\tan(\alpha)}{1-\tan(\alpha)}-\frac{1-\tan(\alpha)}{1+\tan(\alpha)}$$

$$=\frac{(1+\tan(\alpha))^2-(1-\tan(\alpha))^2}{1-\tan^2(\alpha)}$$

$$=\frac{4\tan(\alpha)}{1-\tan^2(\alpha)}$$

$$\Rightarrow \tan\left(\frac{\pi}{4}+\alpha\right)-\tan\left(\frac{\pi}{4}-\alpha\right)=2\tan(2\alpha)$$

Choice (1) is the answer.

4.64. From trigonometry, we know that:

$$\tan\left(\frac{\pi}{4}\right)=1$$

$$\tan(\alpha-\beta)=\frac{\tan(\alpha)-\tan(\beta)}{1+\tan(\alpha)\tan(\beta)}$$

$$\tan(2\alpha)=\frac{2\tan(\alpha)}{1-\tan^2(\alpha)}$$

Moreover, based on the information given in the problem, we have:

$$\tan\left(\frac{\pi}{4}-\alpha\right)=\frac{1}{5} \Rightarrow \frac{\tan\left(\frac{\pi}{4}\right)-\tan(\alpha)}{1+\tan\left(\frac{\pi}{4}\right)\tan(\alpha)}=\frac{1-\tan(\alpha)}{1+\tan(\alpha)}=\frac{1}{5}$$

$$\Rightarrow 5-5\tan(\alpha)=1+\tan(\alpha) \Rightarrow \tan(\alpha)=\frac{2}{3}$$

$$\Rightarrow \tan(2\alpha)=\frac{2\tan(\alpha)}{1-\tan^2(\alpha)}=\frac{2\times\frac{2}{3}}{1-\left(\frac{2}{3}\right)^2}=\frac{12}{5}$$

$$\Rightarrow \tan(2\alpha)=2.4$$

Choice (3) is the answer.

4.65. From trigonometry, we know that:

$$\tan(2\alpha)=\frac{2\tan(\alpha)}{1-\tan^2(\alpha)}$$

$$\tan(\alpha-\beta)=\frac{\tan(\alpha)-\tan(\beta)}{1+\tan(\alpha)\tan(\beta)}$$

Moreover, based on the information given in the problem, we have:

$$\tan(\alpha) = 2$$

$$\tan(\beta) = \frac{1}{3}$$

Therefore:

$$\tan(2\alpha) = \frac{2\tan(\alpha)}{1 - \tan^2(\alpha)} \Rightarrow \tan(2\alpha) = \frac{2 \times 2}{1 - 2^2} = -\frac{4}{3}$$

$$\tan(2\alpha - \beta) = \frac{\tan(2\alpha) - \tan(\beta)}{1 + \tan(2\alpha)\tan(\beta)}$$

$$\Rightarrow \tan(2\alpha - \beta) = \frac{-\frac{4}{3} - \frac{1}{3}}{1 + \left(-\frac{4}{3}\right)\left(\frac{1}{3}\right)} = \frac{-\frac{5}{3}}{\frac{5}{9}}$$

$$\Rightarrow \tan(2\alpha - \beta) = -3$$

Choice (1) is the answer.

4.66. From trigonometry, we know that:

$$\cos(\pi - x) = -\cos(x)$$

$$\cos(\alpha) = \cos(\alpha_0) \Rightarrow \alpha = 2k\pi \pm \alpha_0, \forall k \in \mathbb{Z}$$

Moreover, based on the information given in the problem, we have:

$$\cos(x) \neq 0$$

Therefore:

$$\cos(3x) + \cos(x) = 0 \Rightarrow \cos(3x) = -\cos(x) \Rightarrow \cos(3x) = \cos(\pi - x)$$

$$\Rightarrow 3x = 2k\pi \pm (\pi - x) \Rightarrow \begin{cases} 3x = 2k\pi + \pi - x \Rightarrow 4x = 2k\pi + \pi \\ 3x = 2k\pi - \pi + x \Rightarrow 2x = 2k\pi - \pi \end{cases} \Rightarrow \begin{cases} x = \frac{k\pi}{2} + \frac{\pi}{4} \\ x = k\pi - \frac{\pi}{2} \end{cases}$$

However:

$$\cos(x) \neq 0 \Rightarrow x = \frac{k\pi}{2} + \frac{\pi}{4}$$

Choice (1) is the answer.

4.67. From trigonometry, we know that:

$$\cot(\alpha) = \tan\left(\frac{\pi}{2} - \alpha\right)$$

$$\tan(\alpha) = \tan(\alpha_0) \Rightarrow \alpha = k\pi + \alpha_0, \forall k \in \mathbb{Z}$$

Therefore:

$$\tan(4x) = \cot(x) \Rightarrow \tan(4x) = \tan\left(\frac{\pi}{2} - x\right)$$

$$\Rightarrow 4x = k\pi + \left(\frac{\pi}{2} - x\right) \Rightarrow 5x = k\pi + \frac{\pi}{2} \Rightarrow x = \frac{k\pi}{5} + \frac{\pi}{10}$$

$$\Rightarrow \begin{cases} k = -1 \Rightarrow x_1 = -\dfrac{\pi}{10} \text{ is not a positive angle} \\ k = 0 \Rightarrow x_2 = \dfrac{\pi}{10} \text{ is an acute angle} \\ k = 1 \Rightarrow x_3 = \dfrac{3\pi}{10} \text{ is an acute angle} \\ k = 2 \Rightarrow x_4 = \dfrac{\pi}{2} \text{ is not an acute angle} \end{cases} \Rightarrow x_2 + x_3 = \frac{2\pi}{5}$$

Choice (1) is the answer.

4.68. From trigonometry, we know that:

$$\sin^2(x) + \cos^2(x) = 1$$

$$\cos(x) = \cos(x_0) \Rightarrow x = 2k\pi \pm x_0$$

Therefore:

$$2\sin^2(x) + 3\cos(x) = 0 \Rightarrow 2\left(1 - \cos^2(x)\right) + 3\cos(x) = 0 \Rightarrow 2\cos^2(x) - 3\cos(x) - 2 = 0$$

$$\Rightarrow \cos^2(x) - \frac{3}{2}\cos(x) - 1 = 0 \Rightarrow \left(\cos(x) + \frac{1}{2}\right)(\cos(x) - 2) = 0$$

$$\Rightarrow \begin{cases} \cos(x) = -\dfrac{1}{2} \Rightarrow x = 2k\pi \pm \dfrac{2\pi}{3} \\ \cos(x) = 2 \Rightarrow \text{not acceptable} \end{cases}$$

Choice (1) is the answer.

4.69. From trigonometry, we know that:

$$\sin^2(x) + \cos^2(x) = 1$$

$$\cos(x) = \cos(x_0) \Rightarrow x = 2k\pi \pm x_0$$

Therefore:

$$2\sin^2(x) = 3\cos(x) \Rightarrow 2\left(1 - \cos^2(x)\right) - 3\cos(x) = 0 \Rightarrow 2\cos^2(x) + 3\cos(x) - 2 = 0$$

$$\Rightarrow \cos^2(x) + \frac{3}{2}\cos(x) - 1 = 0 \Rightarrow \left(\cos(x) - \frac{1}{2}\right)(\cos(x) + 2) = 0$$

$$\Rightarrow \begin{cases} \cos(x) = \dfrac{1}{2} \Rightarrow x = 2k\pi \pm \dfrac{\pi}{3} \\ \cos(x) = -2 \Rightarrow \text{not acceptable} \end{cases}$$

Choice (4) is the answer.

4.70. From trigonometry, we know that:

$$\tan(\alpha) . \cot(\alpha) = 1$$

Now, let us find the intersection point of the lines, as follows:

$$\begin{cases} x\tan(\alpha) - y\cot(\alpha) = 1 \\ x\tan(\alpha) + y\cot(\alpha) = 2 \end{cases} \Rightarrow \begin{cases} 2x\tan(\alpha) = 3 \Rightarrow x = \dfrac{3}{2\tan(\alpha)} \\ 2y\cot(\alpha) = 1 \Rightarrow y = \dfrac{1}{2\cot(\alpha)} \end{cases}$$

$$\Rightarrow xy = \frac{3}{2\tan(\alpha)} \times \frac{1}{2\cot(\alpha)} = \frac{3}{4} \Rightarrow y = \frac{3}{4x}$$

Choice (4) is the answer.

4.71. From trigonometry, we know that:

$$\sin^2(\alpha) + \cos^2(\alpha) = 1$$

Based on the information given in the problem, we have:

$$\begin{cases} x = 2 - 3\sin(\alpha) \Rightarrow \sin(\alpha) = \dfrac{x-2}{-3} \\ y = 1 + 4\cos(\alpha) \Rightarrow \cos(\alpha) = \dfrac{y-1}{4} \end{cases}$$

Therefore:

$$\Rightarrow \frac{(x-2)^2}{9} + \frac{(y-1)^2}{16} = 1$$

which is the equation of an ellipse. Choice (2) is the answer.

4.72. Based on the information given in the problem, we have:

$$\begin{cases} x = 2 - 5\cos(\alpha) \\ y = 4 \end{cases}$$

From trigonometry, we know that:

$$-1 \le \cos(\alpha) \le 1 \Rightarrow -1 \le \frac{2-x}{5} \le 1$$

$$\Rightarrow -5 \leq 2-x \leq 5 \Rightarrow -7 \leq -x \leq 3 \Rightarrow -3 \leq x \leq 7$$

Therefore:

$$\Rightarrow \begin{cases} -3 \leq x \leq 7 \\ y=4 \end{cases}$$

which is the equation of a horizontal line segment. Choice (3) is the answer.

4.73. From trigonometry, we know that the maximum value of cos(.) and sin(.) is one. Therefore, the only solution of the given equation is:

$$\begin{cases} \cos(x-y)=1 \\ \sin(x+y)=1 \end{cases}$$

The common solution of the equations can be calculated as follows:

$$\Rightarrow \begin{cases} x-y=2k\pi \\ x+y=2k\pi + \dfrac{\pi}{2} \end{cases} \xrightarrow{0<x,y<2\pi} \begin{cases} x-y=0 \\ x+y= \dfrac{\pi}{2} \ or \ \dfrac{5\pi}{2} \end{cases} \Rightarrow y= \dfrac{\pi}{4} \ or \ \dfrac{5\pi}{4}$$

Choice (2) is the answer.

4.74. From trigonometry, we know that:

$$\tan(\alpha + \beta) = \frac{\tan(\alpha) + \tan(\beta)}{1 - \tan(\alpha)\tan(\beta)}$$

In addition, we know that the sum and the product of the roots of a quadratic equation in the form of $ax^2 + bx + c = 0$ are $-\frac{b}{a}$ and $\frac{c}{a}$, respectively.

Based on the information given in the problem, we have:

$$\tan^2(x) + (m+2)\tan(x) + 2m - 2 = 0$$

$$\alpha + \beta = \frac{\pi}{4}$$

Therefore:

$$\xrightarrow{\tan(.)} \tan(\alpha + \beta) = 1 \Rightarrow \frac{\tan(\alpha) + \tan(\beta)}{1 - \tan(\alpha)\tan(\beta)} = 1$$

$$\Rightarrow \frac{\text{sum of the roots of the quadratic equation}}{1 - \text{product of the roots of the quadratic equation}} = \frac{\frac{-(m+2)}{1}}{1 - \frac{2m-2}{1}} = \frac{-m-2}{3-2m} = 1$$

$$\Rightarrow -m-2 = 3-2m \Rightarrow m = 5$$

Choice (3) is the answer.

4.75. From trigonometry, we know that:

$$\sin^6(\alpha) + \cos^6(\alpha) = 1 - 3\sin^2(\alpha)\cos^2(\alpha)$$

$$\sin^4(\alpha) + \cos^4(\alpha) = 1 - 2\sin^2(\alpha)\cos^2(\alpha)$$

Therefore:

$$\frac{\sin^6(\alpha) + \cos^6(\alpha) + 3\sin^2(\alpha)\cos^2(\alpha)}{\sin^4(\alpha) + \cos^4(\alpha) + 2\sin^2(\alpha)\cos^2(\alpha)} = \frac{1 - 3\sin^2(\alpha)\cos^2(\alpha) + 3\sin^2(\alpha)\cos^2(\alpha)}{1 - 2\sin^2(\alpha)\cos^2(\alpha) + 2\sin^2(\alpha)\cos^2(\alpha)}$$

$$\Rightarrow \frac{\sin^6(\alpha) + \cos^6(\alpha) + 3\sin^2(\alpha)\cos^2(\alpha)}{\sin^4(\alpha) + \cos^4(\alpha) + 2\sin^2(\alpha)\cos^2(\alpha)} = 1$$

Choice (4) is the answer.

4.76. From trigonometry, we know that:

$$\sin(135^\circ) = \sin(180^\circ - 45^\circ) = \sin(45^\circ)$$

$$\cos(210^\circ) = \cos(180^\circ + 30^\circ) = -\cos(30^\circ)$$

$$\cos(135^\circ) = \cos(180^\circ - 45^\circ) = -\cos(45^\circ)$$

$$\sin(420^\circ) = \sin(360^\circ + 60^\circ) = \sin(60^\circ)$$

$$\tan(210^\circ) = \tan(180^\circ + 30^\circ) = \tan(30^\circ)$$

$$\cot(420^\circ) = \cot(360^\circ + 60^\circ) = \cot(60^\circ)$$

$$\cot(120^\circ) = \cot(180^\circ - 60^\circ) = -\cot(60^\circ)$$

$$\tan(330^\circ) = \tan(360^\circ - 30^\circ) = -\tan(30^\circ)$$

Therefore:

$$\frac{\sin(45^\circ)\left(-\cos(30^\circ)\right) + \left(-\cos(45^\circ)\right)\sin(60^\circ)}{\tan(30^\circ)\cot(60^\circ) + \left(-\cot(60^\circ)\right)\left(-\tan(30^\circ)\right)} = \frac{\frac{\sqrt{2}}{2} \times \left(\frac{-\sqrt{3}}{2}\right) + \left(-\frac{\sqrt{2}}{2}\right)\frac{\sqrt{3}}{2}}{\frac{\sqrt{3}}{3} \times \frac{\sqrt{3}}{3} + \left(-\frac{\sqrt{3}}{3}\right)\left(-\frac{\sqrt{3}}{3}\right)}$$

$$\Rightarrow \frac{\sin(45^\circ)\left(-\cos(30^\circ)\right) + \left(-\cos(45^\circ)\right)\sin(60^\circ)}{\tan(30^\circ)\cot(60^\circ) + \left(-\cot(60^\circ)\right)\left(-\tan(30^\circ)\right)} = \frac{-3\sqrt{6}}{4}$$

Choice (2) is the answer.

4.77. From trigonometry, we know that:

$$\cot(x+y) = \frac{\cot(x)\cot(y) - 1}{\cot(x) + \cot(y)}$$

Based on the information given in the problem, we have:

$$x + y = k\pi + \frac{\pi}{4} \xrightarrow{k=0} x + y = \frac{\pi}{4} \xrightarrow{\cot(.)} \frac{\cot(x)\cot(y) - 1}{\cot(x) + \cot(y)} = 1$$

$$\Rightarrow 1 + \cot(x) + \cot(y) = \cot(x)\cot(y) \qquad (1)$$

On the other hand, we can write:

$$(1 + \cot(x))(1 + \cot(y)) = (1 + \cot(x) + \cot(y)) + \cot(x)\cot(y) \qquad (2)$$

Solving (1) and (2):

$$(1 + \cot(x))(1 + \cot(y)) = \cot(x)\cot(y) + \cot(x)\cot(y)$$

$$\Rightarrow (1 + \cot(x))(1 + \cot(y)) = 2\cot(x)\cot(y)$$

Choice (4) is the answer.

4.78. From trigonometry, we know that:

$$\tan\left(\frac{3\pi}{2} - x\right) = \cot(x)$$

$$\cos\left(\frac{4\pi}{3}\right) = -\cos\left(\frac{\pi}{3}\right) = -\frac{1}{2}$$

$$\tan(x)\cot(x) = 1$$

$$\cot(x) = \frac{\cos(x)}{\sin(x)}$$

$$\cos(x) = \cos(x_0) \Rightarrow x = 2k\pi \pm x_0$$

Therefore:

$$(\sin(x) - \tan(x))\tan\left(\frac{3\pi}{2} - x\right) = \cos\left(\frac{4\pi}{3}\right) \Rightarrow (\sin(x) - \tan(x))\cot(x) = -\frac{1}{2}$$

$$\Rightarrow \sin(x)\cot(x) - \tan(x)\cot(x) = -\frac{1}{2} \Rightarrow \cos(x) - 1 = -\frac{1}{2} \Rightarrow \cos(x) = \frac{1}{2}$$

$$\Rightarrow x = 2k\pi \pm \frac{\pi}{3}$$

Choice (3) is the answer.

4.79. From trigonometry, we know that:

$$\sin(\alpha + \beta) = \sin(\alpha)\cos(\beta) + \sin(\alpha)\cos(\beta)$$

$$\cos(\alpha + \beta) = \cos(\alpha)\cos(\beta) - \sin(\alpha)\sin(\beta)$$

$$\tan(x) = \frac{\sin(x)}{\cos(x)}$$

$$\tan(x) = \tan(x_0) \Rightarrow x = k\pi + x_0$$

Therefore:

$$\sin(2x)(\sin(x) + \cos(x)) = \cos(2x)(\cos(x) - \sin(x))$$

$$\Rightarrow \ \sin(2x)\sin(x) + \sin(2x)\cos(x) = \cos(2x)\cos(x) - \cos(2x)\sin(x)$$

$$\Rightarrow \ \sin(2x)\cos(x) + \cos(2x)\sin(x) = \cos(2x)\cos(x) - \sin(2x)\sin(x)$$

$$\Rightarrow \ \sin(2x + x) = \cos(2x + x) \Rightarrow \sin(3x) = \cos(3x) \xrightarrow{\ \times \frac{1}{\cos(3x)} \ \& \ \cos(3x) \neq 0\ } \tan(3x) = 1$$

$$\Rightarrow 3x = k\pi + \frac{\pi}{4} \Rightarrow x = \frac{k\pi}{3} + \frac{\pi}{12} \xrightarrow{\ k=0,1,2 \ \& \ x \in [0,\pi]\ } x_1 = \frac{\pi}{12}, x_2 = \frac{5\pi}{12}, x_3 = \frac{9\pi}{12}$$

$$\Rightarrow x_1 + x_2 + x_3 = \frac{5\pi}{4}$$

Choice (2) is the answer.

4.80. From trigonometry, we know that:

$$\sin\left(\frac{5\pi}{2} + x\right) = \sin\left(\frac{\pi}{2} + x\right) = \cos(x)$$

$$\sin(\alpha + \beta) = \sin(\alpha)\cos(\beta) + \sin(\alpha)\cos(\beta)$$

$$\sin\left(\frac{\pi}{4}\right) = \cos\left(\frac{\pi}{4}\right) = \frac{\sqrt{2}}{2}$$

$$\sin(x) = \sin(x_0) \Rightarrow \begin{cases} x = 2k\pi + x_0 \\ x = 2k\pi + \pi - x_0 \end{cases}$$

Therefore:

$$\Rightarrow \sqrt{2}\sin\left(\frac{\pi}{4} - x\right) = 1 + \cos(x) \Rightarrow \sqrt{2}\left(\sin\left(\frac{\pi}{4}\right)\cos(x) - \cos\left(\frac{\pi}{4}\right)\sin(x)\right) = 1 + \cos(x)$$

$$\Rightarrow \cos(x) - \sin(x) = 1 + \cos(x) \Rightarrow \sin(x) = -1 \Rightarrow \begin{cases} x = 2k\pi + \left(-\dfrac{\pi}{2}\right) \\ x = 2k\pi + \pi - \left(-\dfrac{\pi}{2}\right) \end{cases}$$

$$\Rightarrow \begin{cases} x = 2k\pi - \dfrac{\pi}{2} \\ x = 2k\pi + \dfrac{3\pi}{2} \end{cases}$$

$$\Rightarrow x = 2k\pi - \dfrac{\pi}{2}$$

Choice (3) is the answer.

4.81. From trigonometry, we know that:

$$\tan\left(\frac{\pi}{3}\right) = \sqrt{3}$$

$$\tan(x) = \frac{\sin(x)}{\cos(x)}$$

$$\cos\left(\frac{\pi}{3}\right) = \frac{1}{2}$$

$$\cos(\alpha - \beta) = \cos(\alpha)\cos(\beta) + \sin(\alpha)\sin(\beta)$$

$$\cos(x) = \cos(x_0) \Rightarrow x = 2k\pi \pm x_0$$

Therefore:

$$\cos(2x) + \sqrt{3}\sin(2x) = 1 \Rightarrow \cos(2x) + \tan\left(\frac{\pi}{3}\right)\sin(2x) = 1 \Rightarrow \cos(2x) + \frac{\sin\left(\frac{\pi}{3}\right)}{\cos\left(\frac{\pi}{3}\right)}\sin(2x) = 1$$

$$\Rightarrow \cos(2x)\cos\left(\frac{\pi}{3}\right) + \sin\left(\frac{\pi}{3}\right)\sin(2x) = \cos\left(\frac{\pi}{3}\right) \Rightarrow \cos\left(2x - \frac{\pi}{3}\right) = \cos\left(\frac{\pi}{3}\right)$$

$$\Rightarrow 2x - \frac{\pi}{3} = 2k\pi \pm \frac{\pi}{3} \Rightarrow \begin{cases} 2x - \dfrac{\pi}{3} = 2k\pi + \dfrac{\pi}{3} \Rightarrow x = k\pi + \dfrac{\pi}{3} \\ 2x - \dfrac{\pi}{3} = 2k\pi - \dfrac{\pi}{3} \Rightarrow x = k\pi \end{cases}$$

Choice (4) is the answer.

4.82. From trigonometry, we know that:

$$\sin a \sin b = \frac{1}{2}[\cos(a - b) - \cos(a + b)]$$

$$\sin a \cos b = \frac{1}{2}[\sin(a + b) + \sin(a - b)]$$

$$\cos a + \cos b = 2\cos\left(\frac{a + b}{2}\right)\cos\left(\frac{a - b}{2}\right)$$

$$\sin a - \sin b = 2 \cos\left(\frac{a+b}{2}\right) \sin\left(\frac{a-b}{2}\right)$$

Therefore:

$$\frac{\cos 3\alpha + \sin\alpha \sin 2\alpha}{\sin 3\alpha - \sin 2\alpha \cos\alpha} \times \frac{\sin\alpha}{\cos\alpha} = \frac{\cos 3\alpha + \frac{1}{2}[\cos\alpha - \cos 3\alpha]}{\sin 3\alpha - \frac{1}{2}[\sin 3\alpha + \sin\alpha]} \times \frac{\sin\alpha}{\cos\alpha}$$

$$= \frac{\frac{1}{2}\cos 3\alpha + \frac{1}{2}\cos\alpha}{\frac{1}{2}\sin 3\alpha - \frac{1}{2}\sin\alpha} \times \frac{\sin\alpha}{\cos\alpha} = \frac{\cos 3\alpha + \cos\alpha}{\sin 3\alpha - \sin\alpha} \times \frac{\sin\alpha}{\cos\alpha} = \frac{2\cos 2\alpha \cos\alpha}{2\cos 2\alpha \sin\alpha} \times \frac{\sin\alpha}{\cos\alpha}$$

$$\Rightarrow \frac{\cos 3\alpha + \sin\alpha \sin 2\alpha}{\sin 3\alpha - \sin 2\alpha \cos\alpha} \times \frac{\sin\alpha}{\cos\alpha} = 1$$

Choice (3) is the answer.

References

1. Rahmani-Andebili, M. (2021). Calculus – Practice Problems, Methods, and Solutions, Springer Nature, 2021.
2. Rahmani-Andebili, M. (2021). Precalculus – Practice Problems, Methods, and Solutions, Springer Nature, 2021.

Abstract

In this chapter, the basic and advanced problems of limits and continuities are presented. The subjects include limits by direct substitution, limits by factoring, limits by rationalization, limits at infinity, trigonometric limits, limits of absolute value functions, limits involving Euler's number, limits by L'Hopital's rule, application of Taylor series in limits, and limits and continuity. To help students study the chapter in the most efficient way, the problems are categorized in different levels based on their difficulty levels (easy, normal, and hard) and calculation amounts (small, normal, and large). Moreover, the problems are ordered from the easiest problem with the smallest computations to the most difficult problems with the largest calculations.

5.1. Determine the continuity status of the following function [1, 2]:

$$f(x) = \begin{cases} 10|x| & x \neq 0 \\ 0 & x = 0 \end{cases}$$

Difficulty level   ● Easy   ○ Normal   ○ Hard
Calculation amount   ● Small   ○ Normal   ○ Large
1) It is continuous everywhere except from the right-hand side of $x = 0$.
2) It is continuous everywhere except from the left-hand side of $x = 0$.
3) It is continuous everywhere.
4) It is continuous everywhere except at $x = 0$.

5.2. What is the continuity status of the function below?

$$f(x) = \begin{cases} |x| & x \neq 0 \\ 1 & x = 0 \end{cases}$$

Difficulty level   ● Easy   ○ Normal   ○ Hard
Calculation amount   ● Small   ○ Normal   ○ Large
1) It is continuous everywhere except from the right-hand side of $x = 0$.
2) It is continuous everywhere except from the left-hand side of $x = 0$.
3) It is continuous everywhere.
4) It is continuous everywhere except at $x = 0$.

5.3. Calculate the value of k if the function below is continuous at $x = 2$.

$$f(x) = \begin{cases} (x+2)[-x] & x < 2 \\ x + k & x \geq 2 \end{cases}$$

Difficulty level   ● Easy   ○ Normal   ○ Hard
Calculation amount ● Small   ○ Normal   ○ Large
1) 10
2) −10
3) 6
4) −8

5.4. For which value of the parameter of "a" the function below is continuous at $= -2$?

$$f(x) = \begin{cases} |x|[x] + a & x < -2 \\ |x| + [x] & x \geq -2 \end{cases}$$

Difficulty level   ● Easy   ○ Normal   ○ Hard
Calculation amount ● Small   ○ Normal   ○ Large
1) 0
2) 2
3) 3
4) 6

5.5. Calculate the value of the following limit:

$$\lim_{x \to (-1)^+} \frac{[x] + 1}{x^2 - 1}$$

Difficulty level   ● Easy   ○ Normal   ○ Hard
Calculation amount ● Small   ○ Normal   ○ Large
1) $-\dfrac{1}{2}$
2) 0
3) $\dfrac{1}{2}$
4) ∞

5.6. Calculate the limit of the following function if $x \to 2^+$.

$$f(x) = \frac{x + 4}{[-x] - 3}$$

Difficulty level   ● Easy   ○ Normal   ○ Hard
Calculation amount ● Small   ○ Normal   ○ Large
1) 1
2) −1
3) 2
4) −2

5.7. Determine the value of the following limit:

$$\lim_{x \to 0^-} \frac{x+2}{[x]}$$

1) 2
2) −2
3) 1
4) −1

5.8. Determine the value of the limit below.

$$\lim_{x \to -\infty} \frac{[x] + 3x}{[x] - 3x}$$

1) 2
2) −2
3) 4
4) −4

5.9. Calculate the limit of the function below if $x \to 0$.

$$f(x) = \frac{x + \sqrt[3]{x}}{x - \sqrt[3]{x}}$$

1) 0
2) 1
3) −1
4) ∞

5.10. Calculate the value of the following limit:

$$\lim_{x \to 0^-} \frac{[x]}{x}$$

1) ∞
2) −∞
3) 0
4) 1

5.11. Determine the limit of the function below if $x \rightarrow 0^{+}$.

$$f(x) = \frac{(x^2 - 1)\sqrt{x}}{(x\sqrt{x} + 1)x}$$

Difficulty level   ● Easy   ○ Normal   ○ Hard
Calculation amount   ● Small   ○ Normal   ○ Large
1) ∞
2) $-\infty$
3) 0
4) -1

5.12. Determine the value of the following limit:

$$\lim_{x \rightarrow 0^{+}} \left(\frac{1}{x} - \frac{1}{x^3} \right)$$

Difficulty level   ● Easy   ○ Normal   ○ Hard
Calculation amount   ● Small   ○ Normal   ○ Large
1) ∞
2) $-\infty$
3) 0
4) 1

5.13. For the function below, calculate the value of $\lim\limits_{x \rightarrow 1^{+}} f(x) - \lim\limits_{x \rightarrow 1} f(x)$.

$$f(x) = \frac{2x}{[2x] + 2}$$

Difficulty level   ● Easy   ○ Normal   ○ Hard
Calculation amount   ● Small   ○ Normal   ○ Large
1) $-\infty$
2) $-\dfrac{1}{6}$
3) $\dfrac{2}{3}$
4) ∞

5.14. Calculate the value of $\lim\limits_{x \rightarrow 2^{+}} ([x] - 2)[x]$.
Difficulty level   ● Easy   ○ Normal   ○ Hard
Calculation amount   ● Small   ○ Normal   ○ Large
1) -2
2) -1
3) 0
4) 1

5.15. Calculate the limit of the following function if $x \rightarrow 4^{-}$:

$$f(x) = \frac{[x] - 4}{x^2 - 16}$$

1) 0
2) $\dfrac{1}{8}$
3) ∞
4) $-\infty$

5.16. Determine the value of the limit below.

$$\lim_{x \to 1^-} \frac{1 - x^3}{\mathrm{arc}(\cos(x))}$$

1) 1
2) -1
3) 0
4) -3

5.17. Calculate the value of the limit below.

$$\lim_{x \to 0} \frac{\tan(x) - \tan(3x) + \tan(2x)}{x^3}$$

1) -6
2) 6
3) 10
4) -10

5.18. Calculate the value of the following limit:

$$\lim_{x \to 3} \frac{9 - x^2}{2 - \sqrt{x + 1}}$$

1) 6
2) 12
3) 18
4) 24

5.19. Determine the limit of the function below if $x \to +\infty$.

$$f(x) = \frac{\sin(x)}{x}$$

1) Undefined
2) 0
3) 1
4) ∞

5.20. Determine the value of the limit below.

$$\lim_{x \to 0} \frac{[x^2] - x^2}{x \tan(x)}$$

Difficulty level   ○ Easy   ● Normal   ○ Hard
Calculation amount   ● Small   ○ Normal   ○ Large
1) 1
2) -1
3) 2
4) -2

5.21. Determine the value of the following limit:

$$\lim_{x \to +\infty} x \sin\left(\frac{1}{x}\right)$$

Difficulty level   ○ Easy   ● Normal   ○ Hard
Calculation amount   ● Small   ○ Normal   ○ Large
1) 1
2) -1
3) 0
4) Undefined

5.22. Calculate the value of the following limit:

$$\lim_{x \to -\infty} \left(\frac{x^2 + x - 1}{-3x + 4\sqrt{-x}} \right)$$

Difficulty level   ○ Easy   ● Normal   ○ Hard
Calculation amount   ● Small   ○ Normal   ○ Large
1) $\frac{1}{3}$
2) $-\frac{1}{3}$
3) ∞
4) $-\infty$

5.23. Calculate the value of the limit below.

$$\lim_{x \to 0^+} \frac{(x+1)\sqrt{x}}{x^2 - x}$$

Difficulty level   ○ Easy   ● Normal   ○ Hard
Calculation amount   ● Small   ○ Normal   ○ Large
1) 0

2) -1
3) ∞
4) $-\infty$

5.24. Determine the limit of the following function if $x \rightarrow +\infty$:

$$f(x) = \frac{x}{x - 1 + \sqrt{x^2 + x - 1}}$$

Difficulty level   ○ Easy   ● Normal   ○ Hard
Calculation amount   ● Small   ○ Normal   ○ Large
1) ∞
2) 0
3) $-\dfrac{1}{2}$
4) $\dfrac{1}{2}$

5.25. Calculate the value of $\lim\limits_{x \to 0} x \cot(x)$.

Difficulty level   ○ Easy   ● Normal   ○ Hard
Calculation amount   ● Small   ○ Normal   ○ Large
1) 0
2) ∞
3) 1
4) 2

5.26. For what value of "a" the following function has a definite limit at $x = 1$?

$$f(x) = \begin{cases} x^2 + ax & x > 1 \\ x - 3 & x < 1 \end{cases}$$

Difficulty level   ○ Easy   ● Normal   ○ Hard
Calculation amount   ● Small   ○ Normal   ○ Large
1) 0
2) 3
3) -3
4) -2

5.27. Determine the value of the limit below.

$$\lim_{x \to 2^-} \frac{|x^3 - 8|}{x - \sqrt{2x}}$$

Difficulty level   ○ Easy   ● Normal   ○ Hard
Calculation amount   ● Small   ○ Normal   ○ Large
1) -24
2) -16
3) 16
4) 24

5.28. Calculate the value of the limit below.

$$\lim_{x \to 0^-} \frac{[x] + x}{[-x] + x}$$

Difficulty level   ○ Easy   ● Normal   ○ Hard
Calculation amount   ● Small   ○ Normal   ○ Large
1) $+\infty$
2) $-\infty$
3) 1
4) -1

5.29. Determine the value of the limit below.

$$\lim_{x \to 0} \frac{\sin(3x) + \sin(7x)}{3x + \tan(2x)}$$

Difficulty level   ○ Easy   ● Normal   ○ Hard
Calculation amount   ● Small   ○ Normal   ○ Large
1) 1
2) 2
3) -1
4) -2

5.30. Calculate the value of the following limit:

$$\lim_{x \to 0} \frac{\sqrt{x + 3} - \sqrt{3}}{x}$$

Difficulty level   ○ Easy   ● Normal   ○ Hard
Calculation amount   ● Small   ○ Normal   ○ Large
1) $\dfrac{\sqrt{3}}{3}$

2) $\dfrac{\sqrt{3}}{6}$

3) $\dfrac{\sqrt{3}}{2}$

4) $\dfrac{\sqrt{3}}{9}$

5.31. Calculate the value of the following limit:

$$\lim_{x \to 0} \frac{1 - \cos(x)}{\sin(x)}$$

Difficulty level   ○ Easy   ● Normal   ○ Hard
Calculation amount   ● Small   ○ Normal   ○ Large
1) 0
2) 1
3) -1

4) $\sqrt{2}$

5.32. Calculate the value of the limit below.

$$\lim_{x \to 0} \frac{5x - \sin(x)}{2x + \cos(x) - 1}$$

Difficulty level   ○ Easy   ● Normal   ○ Hard
Calculation amount   ● Small   ○ Normal   ○ Large
1) 1
2) 2
3) −1
4) −2

5.33. Determine the value of the limit below.

$$\lim_{x \to 2^-} \frac{x^3 - 8}{|x - 2|} + 5x$$

Difficulty level   ○ Easy   ● Normal   ○ Hard
Calculation amount   ● Small   ○ Normal   ○ Large
1) 2
2) −2
3) 1
4) −1

5.34. Calculate the value of the limit below.

$$\lim_{x \to \left(\frac{\pi}{2}\right)^+} \frac{\sin(x) + \cos(x)}{\cos(x)}$$

Difficulty level   ○ Easy   ● Normal   ○ Hard
Calculation amount   ● Small   ○ Normal   ○ Large
1) ∞
2) $-\infty$
3) 0
4) 1

5.35. Calculate the value of the following limit:

$$\lim_{x \to 0} \frac{3x^4 + 2x^3}{(\arc(\sin(x)))^3}$$

Difficulty level   ○ Easy   ● Normal   ○ Hard
Calculation amount   ● Small   ○ Normal   ○ Large
1) 1
2) 2
3) 0
4) ∞

5.36. Determine the value of the limit below.

$$\lim_{x \to -\infty} \left[\frac{2}{x+1} \right] x$$

Difficulty level   ○ Easy   ● Normal   ○ Hard
Calculation amount   ● Small   ○ Normal   ○ Large
1) ∞
2) 2
3) 0
4) −∞

5.37. Calculate the value of the following limit:

$$\lim_{x \to 3^+} \frac{x-4}{\sqrt{x^2 - 4x + 3}}$$

Difficulty level   ○ Easy   ● Normal   ○ Hard
Calculation amount   ● Small   ○ Normal   ○ Large
1) ∞
2) −∞
3) 1
4) −1

5.38. Calculate the value of $\lim_{x \to -\infty} \left(x + \sqrt{x^2 + 4x - 10} \right)$.

Difficulty level   ○ Easy   ● Normal   ○ Hard
Calculation amount   ● Small   ○ Normal   ○ Large
1) 2
2) −2
3) ∞
4) −∞

5.39. Determine the value of the limit below.

$$\lim_{x \to 2} \frac{4 - x^2}{6 - 2\sqrt{x^2 + 5}}$$

Difficulty level   ○ Easy   ● Normal   ○ Hard
Calculation amount   ● Small   ○ Normal   ○ Large
1) 0
2) 2
3) 3
4) 1

5.40. Calculate the value of the following limit:

$$\lim_{x \to 0} \frac{\sin(2x)}{\sqrt{x+1} - 1}$$

Difficulty level   ○ Easy   ● Normal   ○ Hard
Calculation amount   ● Small   ○ Normal   ○ Large
1) 2
2) 4

3) 3

4) 1

5.41. Calculate the limit of $\sqrt{x^4 + 2x^2 + x} - x^2$ if $x \to -\infty$.

Difficulty level   ○ Easy   ● Normal   ○ Hard

Calculation amount   ○ Small   ● Normal   ○ Large

1) 1

2) $+\infty$

3) 0

4) $-\infty$

5.42. Calculate the value of the limit below.

$$\lim_{x \to -3} \frac{|x^2 - 9|}{x + 3}$$

Difficulty level   ○ Easy   ● Normal   ○ Hard

Calculation amount   ○ Small   ● Normal   ○ Large

1) 6

2) -6

3) 3

4) Undefined

5.43. Calculate the value of the following limit:

$$\lim_{x \to \frac{1}{2}} \frac{\tan \frac{\pi x}{2} - 1}{\cos(\pi x)}$$

Difficulty level   ○ Easy   ● Normal   ○ Hard

Calculation amount   ○ Small   ● Normal   ○ Large

1) 1

2) -1

3) 2

4) -2

5.44. Calculate the limit of $\sqrt{x + 5} - \sqrt{x + 1}$ if $x \to \infty$.

Difficulty level   ○ Easy   ● Normal   ○ Hard

Calculation amount   ○ Small   ● Normal   ○ Large

1) 4

2) 2

3) 0

4) ∞

5.45. Calculate the value of the following limit:

$$\lim_{x \to \frac{\pi}{2}} \frac{\tan(2x) \cos(x)}{1 + \cos(2x)}$$

Difficulty level   ○ Easy   ● Normal   ○ Hard

Calculation amount   ○ Small   ● Normal   ○ Large

1) 1

2) $\frac{1}{2}$

3) -1

4) $-\dfrac{1}{2}$

5.46. Determine the value of the following limit:

$$\lim_{x \to 0^-} \frac{\tan(2x)}{\sqrt{1 - \cos(x)}}$$

Difficulty level   ○ Easy   ● Normal   ○ Hard
Calculation amount   ○ Small   ● Normal   ○ Large
1) $-2\sqrt{2}$
2) $-\sqrt{2}$
3) $\sqrt{2}$
4) $2\sqrt{2}$

5.47. Calculate the value of $\lim\limits_{x \to -\infty} \left(\sqrt[3]{n + 1000} - \sqrt[3]{n - 20}\right)$.

Difficulty level   ○ Easy   ● Normal   ○ Hard
Calculation amount   ○ Small   ● Normal   ○ Large
1) 2
2) 0
3) 10
4) 20

5.48. Calculate the value of the limit below.

$$\lim_{x \to \pi^+} \frac{\sin(\pi \sin(x)) \sin\left(\frac{x}{2}\right)}{\sqrt{1 + \cos(x)}}$$

Difficulty level   ○ Easy   ● Normal   ○ Hard
Calculation amount   ○ Small   ● Normal   ○ Large
1) $-\pi\sqrt{2}$
2) -2π
3) π^2
4) $\pi^2\sqrt{2}$

5.49. Calculate the value of the following limit.

$$\lim_{x \to 0} \frac{\sqrt[3]{1 + x^2} - \sqrt[4]{1 - 2x}}{2x^2 + 2x}$$

Difficulty level   ○ Easy   ● Normal   ○ Hard
Calculation amount   ○ Small   ● Normal   ○ Large
1) $\dfrac{1}{4}$
2) $-\dfrac{1}{4}$
3) 1
4) -1

5.50. Calculate the limit of the function below if $x \to 0$.

$$f(x) = \frac{\sin^2(x) + \sin(x) + \cos^2(x) - \cos(x)}{\sin^2(x) - \sin(x) + \cos^2(x) - \cos(x)}$$

Difficulty level   ○ Easy   ● Normal   ○ Hard
Calculation amount   ○ Small   ● Normal   ○ Large
1) 1
2) −1
3) 2
4) −2

5.51. Determine the value of the following limit:

$$\lim_{x \to 0} \frac{\cos(mx) - \cos(nx)}{x^2}$$

Difficulty level   ○ Easy   ● Normal   ○ Hard
Calculation amount   ○ Small   ● Normal   ○ Large
1) $n^2 + m^2$
2) $n^2 - m^2$
3) $\dfrac{n^2 - m^2}{2}$
4) $\dfrac{n^2 + m^2}{2}$

5.52. Calculate the value of the limit below.

$$\lim_{x \to 0} \frac{\sin(x) - x}{\tan(x) - x}$$

Difficulty level   ○ Easy   ● Normal   ○ Hard
Calculation amount   ○ Small   ● Normal   ○ Large
1) $-\dfrac{1}{2}$
2) $\dfrac{1}{2}$
3) $-\dfrac{1}{4}$
4) $\dfrac{1}{4}$

5.53. Calculate the value of the following limit:

$$\lim_{x \to \pi} \frac{1 + \cos^3(x)}{1 - \cos^2(x)}$$

Difficulty level   ○ Easy   ● Normal   ○ Hard
Calculation amount   ○ Small   ● Normal   ○ Large
1) $\dfrac{3}{2}$
2) $-\dfrac{3}{2}$

3) 3

4) −3

5.54. For the following function, we have $f(0) = 0$. Which one of the choices below is correct about the continuity of the function at $x = 0$?

$$f(x) = x(-1)^{\left[\frac{1}{x}\right]}, \quad x \in \mathbb{R} - \{0\}$$

Difficulty level   ○ Easy   ○ Normal   ● Hard

Calculation amount   ● Small   ○ Normal   ○ Large

1) The function has only right-hand side continuity at $x = 0$.

2) The function is only left continuous at $x = 0$.

3) The function is continuous at $x = 0$.

4) The function is not continuous at $x = 0$.

5.55. Calculate the limit of the function below if $n \to +\infty$.

$$f(n) = \frac{3n^2}{\sqrt{5^n}}$$

Difficulty level   ○ Easy   ○ Normal   ● Hard

Calculation amount   ● Small   ○ Normal   ○ Large

1) 0

2) 1

3) ∞

4) $\dfrac{3}{\sqrt{5}}$

5.56. Determine the value of the limit below.

$$\lim_{x \to 0} \frac{x^3 - \sin(x)(1 - \cos(x))}{x^3}$$

Difficulty level   ○ Easy   ○ Normal   ● Hard

Calculation amount   ○ Small   ● Normal   ○ Large

1) 0

2) $\dfrac{1}{2}$

3) 1

4) −1

5.57. Calculate the value of the following limit:

$$\lim_{x \to 1^-} \frac{\arc(\cos x)}{\sqrt{1 - x}}$$

Difficulty level   ○ Easy   ○ Normal   ● Hard

Calculation amount   ○ Small   ● Normal   ○ Large

1) $\sqrt{2}$

2) $-\sqrt{2}$

3) $-\dfrac{\sqrt{2}}{2}$

4) $\dfrac{\sqrt{2}}{2}$

5.58. Determine the value of n in the following equation:

$$\lim_{x \to 1} (x^2 - 1) \cot(x^n - 1) = \frac{1}{2}$$

Difficulty level   ○ Easy   ○ Normal   ● Hard
Calculation amount   ○ Small   ● Normal   ○ Large
1) 8
2) 4
3) $\dfrac{1}{8}$
4) $\dfrac{1}{4}$

5.59. Calculate the limit of $\sin(4x)(\cot(2x) - \cot(x))$ if $x \to 0$.
Difficulty level   ○ Easy   ○ Normal   ● Hard
Calculation amount   ○ Small   ● Normal   ○ Large
1) 4
2) 2
3) -2
4) -4

5.60. Calculate the value of the following limit:

$$\lim_{x \to 0^-} \frac{\sin(x) - x}{\frac{1}{2} \sin(2x) - x \cos(x)}$$

Difficulty level   ○ Easy   ○ Normal   ● Hard
Calculation amount   ○ Small   ● Normal   ○ Large
1) 0
2) ∞
3) 1
4) $-\infty$

5.61. Calculate the limit of the following function if $x \to \dfrac{\pi}{4}$:

$$f(x) = \frac{1 - \sqrt[3]{\tan(x)}}{1 - 2\sin^2(x)}$$

Difficulty level   ○ Easy   ○ Normal   ● Hard
Calculation amount   ○ Small   ● Normal   ○ Large
1) $\dfrac{1}{3}$
2) 3

3) $\dfrac{1}{2}$

4) 2

5.62. Calculate the value of the limit below.

$$\lim_{x \to 0} \frac{1 - \cos^3(x)}{\sin(x)\tan(2x)}$$

Difficulty level   ○ Easy   ○ Normal   ● Hard
Calculation amount   ○ Small   ● Normal   ○ Large

1) 4

2) −4

3) $\dfrac{3}{4}$

4) $-\dfrac{3}{4}$

5.63. Determine the value of the following limit:

$$\lim_{x \to 0^+} \frac{1 - \sqrt{\cos(x)}}{1 - \cos(\sqrt{x})}$$

Difficulty level   ○ Easy   ○ Normal   ● Hard
Calculation amount   ○ Small   ○ Normal   ● Large

1) 0

2) 1

3) $\dfrac{1}{2}$

4) 2

References

1. Rahmani-Andebili, M. (2021). Calculus – Practice Problems, Methods, and Solutions, Springer Nature, 2021.
2. Rahmani-Andebili, M. (2021). Precalculus – Practice Problems, Methods, and Solutions, Springer Nature, 2021.

Abstract

In this chapter, the problems of the fifth chapter are fully solved, in detail, step-by-step, and with different methods.

6.1. Based on the information given in the problem, we have [1, 2]:

$$f(x) = \begin{cases} 10|x| & x \neq 0 \\ 0 & x = 0 \end{cases}$$

The function of $10|x|$ is continuous everywhere on real numbers ($\mathbb{R}$); however, the continuity of the $f(x)$ must be checked at $x = 0$.

A function is continuous at the given point of x_0 if:

$$\lim_{x \to x_0^+} f(x) = \lim_{x \to x_0^-} f(x) = f(x_0)$$

As can be seen, for the $f(x)$, we have:

$$\lim_{x \to 0^+} 10|x| = 0$$

$$\lim_{x \to 0^-} 10|x| = 0$$

$$f(0) = 0$$

$$\Rightarrow \lim_{x \to 0^+} 10|x| = \lim_{x \to 0^-} 10|x| = f(0)$$

Therefore, the function is continuous everywhere on $\mathbb{R}$.

Choice (3) is the answer.

6.2. Based on the information given in the problem, we have:

$$f(x) = \begin{cases} |x| & x \neq 0 \\ 1 & x = 0 \end{cases}$$

The function of $|x|$ is continuous everywhere on $\mathbb{R}$; however, the continuity of the $f(x)$ must be checked at $x = 0$.

A function is continuous at the given point of x_0 if:

$$\lim_{x \to x_0^+} f(x) = \lim_{x \to x_0^-} f(x) = f(x_0)$$

As can be seen, for the abovementioned function, we have:

$$\lim_{x \to 0^+} |x| = 0$$

$$\lim_{x \to 0^-} |x| = 0$$

$$f(0) = 1$$

$$\Rightarrow \lim_{x \to 0^+} |x| = \lim_{x \to 0^-} |x| \neq f(0)$$

Therefore, the function is continuous everywhere on $\mathbb{R}$ except at $x = 0$.

Choice (4) is the answer.

6.3. Based on the information given in the problem, the following function is continuous at $x = 2$.

$$f(x) = \begin{cases} (x+2)[-x] & x < 2 \\ x + k & x \geq 2 \end{cases}$$

Therefore, we must have:

$$\lim_{x \to 2^+} f(x) = \lim_{x \to 2^-} f(x) = f(2)$$

For $f(x)$, we have:

$$\Rightarrow \begin{cases} \lim_{x \to 2^+} f(x) = \lim_{x \to 2^+} (x + k) = 2 + k \\ \lim_{x \to 2^-} f(x) = \lim_{x \to 2^-} (x+2)[-x] = (2^- + 2)[-2^-] = -8 \\ f(2) = 2 + k \end{cases}$$

$$\Rightarrow 2 + k = -8$$

$$\Rightarrow k = -10$$

Choice (2) is the answer.

6.4. Based on the information given in the problem, the function is continuous at $x = -2$.

$$f(x) = \begin{cases} |x|[x] + a & x < -2 \\ |x| + [x] & x \geq -2 \end{cases}$$

Therefore, we must have:

$$\lim_{x \to -2^+} f(x) = \lim_{x \to -2^-} f(x) = f(-2)$$

For $f(x)$, we can write:

$$\Rightarrow \begin{cases} \lim\limits_{x \to -2^+} f(x) = \lim\limits_{x \to -2^+} (|x| + [x]) = 2 + (-2) = 0 \\ \lim\limits_{x \to -2^-} f(x) = \lim\limits_{x \to -2^-} |x|[x] + a = (2)(-3) + a = -6 + a \\ f(-2) = |-2| + [-2] = 0 \end{cases}$$

$$\Rightarrow -6 + a = 0$$

$$\Rightarrow a = 6$$

Choice (4) is the answer.

6.5. The problem can be solved as follows:

$$\lim\limits_{x \to (-1)^+} \frac{[x] + 1}{x^2 - 1} = \frac{(-1) + 1}{1^- - 1} = \frac{0}{0^-} = 0$$

Choice (2) is the answer.

6.6. The problem can be solved as follows:

$$\lim\limits_{x \to 2^+} \frac{x + 4}{[-x] - 3} = \frac{2 + 4}{-3 - 3} = -1$$

Choice (2) is the answer.

6.7. The problem can be solved as follows:

$$\lim\limits_{x \to 0^-} \frac{x + 2}{[x]} = \frac{0 + 2}{-1} = -2$$

Choice (2) is the answer.

6.8. The problem can be solved as follows:

$$\lim\limits_{x \to -\infty} \frac{[x] + 3x}{[x] - 3x} = \lim\limits_{x \to -\infty} \frac{4x}{-2x} = \lim\limits_{x \to -\infty} (-2) = -2$$

Choice (2) is the answer.

6.9. The problem can be solved as follows:

$$\lim\limits_{x \to 0} \frac{x + \sqrt[3]{x}}{x - \sqrt[3]{x}} = \lim\limits_{x \to 0} \frac{\sqrt[3]{x}\left(\sqrt[3]{x^2} + 1\right)}{\sqrt[3]{x}\left(\sqrt[3]{x^2} - 1\right)} = \lim\limits_{x \to 0} \frac{\left(\sqrt[3]{x^2} + 1\right)}{\left(\sqrt[3]{x^2} - 1\right)} = -1$$

Choice (3) is the answer.

6.10. The problem can be solved as follows:

$$\lim_{x \to 0} \frac{[x]}{x} = \frac{-1}{0^-} = +\infty$$

Choice (1) is the answer.

6.11. The problem can be solved as follows:

$$\lim_{x \to 0^+} \frac{(x^2-1)\sqrt{x}}{(x\sqrt{x}+1)x} = \lim_{x \to 0^+} \frac{(x^2-1)}{(x\sqrt{x}+1)\sqrt{x}} = \frac{-1}{0^+} = -\infty$$

Choice (2) is the answer.

6.12. The problem can be solved as follows:

$$\lim_{x \to 0^+} \left(\frac{1}{x} - \frac{1}{x^3} \right) = \lim_{x \to 0^+} \left(\frac{x^2-1}{x^3} \right) = \frac{-1}{0^+} = -\infty$$

Choice (2) is the answer.

6.13. The problem can be solved as follows:

$$\lim_{x \to 1^+} f(x) - \lim_{x \to 1^-} f(x) = \lim_{x \to 1^+} \frac{2x}{[2x]+2} - \lim_{x \to 1^-} \frac{2x}{[2x]+2} = \lim_{x \to 1^+} \frac{2x}{2+2} - \lim_{x \to 1^-} \frac{2x}{1+2}$$

$$= \lim_{x \to 1^+} \frac{x}{2} - \lim_{x \to 1^-} \frac{2}{3}x = \frac{1}{2} - \frac{2}{3} = \frac{-1}{6}$$

Choice (2) is the answer.

6.14. The problem can be solved as follows:

$$\lim_{x \to 2^+} ([x]-2)[x] = \lim_{x \to 2^+} (2-2) \times 2 = \lim_{x \to 2^+} 0 = 0$$

Choice (3) is the answer.

6.15. The problem can be solved as follows:

$$\lim_{x \to 4^-} \frac{[x]-4}{x^2-16} = \lim_{x \to 4^-} \frac{3-4}{16^- - 16} = \frac{-1}{0^-} = +\infty$$

Choice (3) is the answer.

6.16. From trigonometry and calculus, we know that:

$$\text{arc}(\cos(1^-)) = 0^+$$

$$\frac{d}{dx} (\text{arc}(\cos x)) = \frac{-1}{\sqrt{1-x^2}}$$

The problem can be solved as follows:

$$\lim_{x \to 1} \frac{1 - x^3}{\arc(\cos(x))} = \frac{0^+}{0^+}$$

$$\xRightarrow{H} \lim_{x \to 1^-} \frac{\frac{d}{dx}(1 - x^3)}{\frac{d}{dx}(\arc(\cos(x)))} = \lim_{x \to 1^-} \frac{-3x^2}{\frac{-1}{\sqrt{1 - x^2}}} = \lim_{x \to 1} 3x^2 \sqrt{1 - x^2} = 0$$

Choice (3) is the answer.

6.17. From application of Taylor series in limit, we know that:

$$\lim_{x \to 0} \tan(x) \equiv x + \frac{x^3}{3}$$

The problem can be solved as follows:

$$\lim_{x \to 0} \frac{\tan(x) - \tan(3x) + \tan(2x)}{x^3} \equiv \lim_{x \to 0} \frac{x + \frac{x^3}{3} - \left(3x + \frac{(3x)^3}{3}\right) + 2x + \frac{(2x)^3}{3}}{x^3}$$

$$= \lim_{x \to 0} \frac{-6x^3}{x^3} = \lim_{x \to 0} (-6) = -6$$

Choice (1) is the answer.

6.18. The problem can be solved as follows:

$$\lim_{x \to 3} \frac{9 - x^2}{2 - \sqrt{x + 1}} = \frac{0}{0}$$

$$\xRightarrow{H} \lim_{x \to 3} \frac{\frac{d}{dx}(9 - x^2)}{\frac{d}{dx}(2 - \sqrt{x + 1})} = \lim_{x \to 3} \frac{-2x}{\frac{-1}{2\sqrt{x+1}}} = \lim_{x \to 3} 4x\sqrt{x + 1} = 24$$

Choice (4) is the answer.

6.19. From trigonometry, we know that:

$$-1 \le \sin(x) \le 1$$

The problem can be solved as follows:

$$\lim_{x \to +\infty} \frac{\sin(x)}{x} = \lim_{x \to +\infty} \sin(x) \times \frac{1}{x} = (\text{Bounded quantity}) \times 0 = 0$$

Choice (2) is the answer.

6.20. From application of Taylor series in limit, we know that:

$$\lim_{x \to 0} \tan(x) \equiv x$$

The problem can be solved as follows:

$$\lim_{x \to 0} \frac{[x^2] - x^2}{x \, \tan(x)} = \lim_{x \to 0} \frac{0 - x^2}{x \, \tan(x)} \equiv \lim_{x \to 0} \frac{-x^2}{x \times x} = \lim_{x \to 0} (-1) = -1$$

Choice (2) is the answer.

6.21. From application of Taylor series in limit, we know that:

$$\lim_{x \to +\infty} \sin\left(\frac{1}{x}\right) \equiv \frac{1}{x}$$

The problem can be solved as follows:

$$\lim_{x \to +\infty} x \sin\left(\frac{1}{x}\right) = \lim_{x \to +\infty} x \times \frac{1}{x} = \lim_{x \to +\infty} 1 = 1$$

Choice (1) is the answer.

6.22. From calculus, we know that:

$$\lim_{x \to \pm\infty} \left(a_m x^m + a_{m-1} x^{m-1} + \ldots + a_2 x^2 + a_1 x + a_0\right) \equiv a_m x^m$$

or:

$$\lim_{x \to \pm\infty} \left(a_m x^m + a_n x^n\right) \equiv a_m x^m \quad if \; m > n$$

Therefore:

$$\lim_{x \to -\infty} \left(\frac{x^2 + x - 1}{-3x + 4\sqrt{-x}}\right) \equiv \lim_{x \to -\infty} \frac{x^2}{-3x} = \lim_{x \to -\infty} \left(-\frac{x}{3}\right) = +\infty$$

Choice (3) is the answer.

6.23. From calculus, we know that:

$$\lim_{x \to 0} \left(a_m x^m + a_{m-1} x^{m-1} + \ldots + a_{m-n} x^{m-n} + a_{m-n-1} x^{m-n-1}\right) \equiv a_{m-n-1} x^{m-n-1}$$

or:

$$\lim_{x \to 0} \left(a_m x^m + a_n x^n\right) \equiv a_n x^n \quad if \; m > n$$

The problem can be solved as follows:

$$\lim_{x \to 0^+} \frac{(x+1)\sqrt{x}}{x^2 - x} = \lim_{x \to 0^+} \frac{x\sqrt{x} + \sqrt{x}}{x^2 - x} \equiv \lim_{x \to 0^+} \frac{\sqrt{x}}{-x} = \lim_{x \to 0^+} \frac{-1}{\sqrt{x}} = -\infty$$

Choice (4) is the answer.

6.24. From calculus, we know that:

$$\lim_{x \to +\infty} \sqrt{x^2 + ax + b} \equiv \left| x + \frac{a}{2} \right|$$

The problem can be solved as follows:

$$\lim_{x \to +\infty} \frac{x}{x - 1 + \sqrt{x^2 + x - 1}} \equiv \lim_{x \to +\infty} \frac{x}{x - 1 + \left| x + \frac{1}{2} \right|} = \lim_{x \to +\infty} \frac{x}{2x - \frac{1}{2}} \equiv \lim_{x \to +\infty} \frac{x}{2x} = \lim_{x \to +\infty} \frac{1}{2} = \frac{1}{2}$$

Choice (4) is the answer.

6.25. From trigonometry, we know that:

$$\cot(x) = \frac{1}{\tan(x)}$$

From application of Taylor series in limit, we know that:

$$\lim_{x \to 0} \tan(x) \equiv x$$

The problem can be solved as follows:

$$\lim_{x \to 0} x \cot(x) = \lim_{x \to 0} \frac{x}{\tan(x)} \equiv \lim_{x \to 0} \frac{x}{x} = 1$$

Choice (3) is the answer.

6.26. As we know, the limit of a function at the point of x_0 exits if:

$$\lim_{x \to x_0^-} f(x) = \lim_{x \to x_0^+} f(x) \Rightarrow \lim_{x \to 1^-} f(x) = \lim_{x \to 1^+} f(x) \tag{1}$$

Therefore:

$$\lim_{x \to 1^-} f(x) = \lim_{x \to 1^-} (x - 3) = 1 - 3 = -2 \tag{2}$$

$$\lim_{x \to 1^+} f(x) = \lim_{x \to 1^+} (x^2 + ax) = 1 + a \tag{3}$$

$$\xrightarrow{Using \ (1), (2), (3)} -2 = 1 + a \Rightarrow a = -3$$

Choice (3) is the answer.

6.27. The problem can be solved as follows:

$$\lim_{x \to 2^-} \frac{|x^3 - 8|}{x - \sqrt{2x}} = \lim_{x \to 2^-} \frac{-(x^3 - 8)}{x - \sqrt{2x}} = \frac{0^+}{0^-}$$

$$\xrightarrow{H} \lim_{x \to 2^-} \frac{\frac{d}{dx}(-(x^3 - 8))}{\frac{d}{dx}(x - \sqrt{2x})} = \lim_{x \to 2^-} \frac{-3x^2}{1 - \frac{1}{\sqrt{2x}}} = \frac{-3 \times 2^2}{1 - \frac{1}{\sqrt{2 \times 2}}} = \frac{-12}{\frac{1}{2}} = -24$$

Choice (1) is the answer.

6.28. The problem can be solved as follows:

$$\lim_{x \to 0} \frac{[x] + x}{[-x] + x} = \lim_{x \to 0^-} \frac{-1 + x}{0 + x} = \frac{-1 + 0^-}{0 + 0^-} = \frac{-1}{0^-} = +\infty$$

Choice (1) is the answer.

6.29. The problem can be solved as follows:

$$\lim_{x \to 0} \frac{\sin(3x) + \sin(7x)}{3x + \tan(2x)} = \frac{0}{0}$$

$$\xrightarrow{H} \lim_{x \to 0} \frac{\frac{d}{dx}(\sin(3x) + \sin(7x))}{\frac{d}{dx}(3x + \tan(2x))} = \lim_{x \to 0} \frac{3\cos(3x) + 7\cos(7x)}{3 + 2(1 + \tan^2(2x))} = \frac{3 + 7}{3 + 2} = 2$$

Choice (2) is the answer.

6.30. The problem can be solved as follows:

$$\lim_{x \to 0} \frac{\sqrt{x + 3} - \sqrt{3}}{x} = \lim_{x \to 0} \frac{\sqrt{x + 3} - \sqrt{3}}{x} \times \frac{\sqrt{x + 3} + \sqrt{3}}{\sqrt{x + 3} + \sqrt{3}} = \lim_{x \to 0} \frac{x + 3 - 3}{x(\sqrt{x + 3} + \sqrt{3})}$$

$$= \lim_{x \to 0} \frac{1}{(\sqrt{x + 3} + \sqrt{3})} = \frac{1}{\sqrt{3} + \sqrt{3}} = \frac{1}{2\sqrt{3}} = \frac{\sqrt{3}}{6}$$

Choice (2) is the answer.

6.31. The problem can be solved as follows:

$$\lim_{x \to 0} \frac{1 - \cos(x)}{\sin(x)} = \frac{0}{0}$$

$$\xrightarrow{H} \lim_{x \to 0} \frac{\frac{d}{dx}(1 - \cos(x))}{\frac{d}{dx}\sin(x)} = \lim_{x \to 0} \frac{\sin(x)}{\cos(x)} = \frac{0}{1} = 0$$

Choice (1) is the answer.

6.32. The problem can be solved as follows:

$$\lim_{x \to 0} \frac{5x - \sin(x)}{2x + \cos(x) - 1} = \frac{0}{0}$$

$$\xrightarrow{H} \lim_{x \to 0} \frac{\frac{d}{dx}(5x - \sin(x))}{\frac{d}{dx}(2x + \cos(x) - 1)} = \lim_{x \to 0} \frac{5 - \cos(x)}{2 - \sin(x)} = \frac{5 - 1}{2 - 0} = 2$$

Choice (2) is the answer.

6.33. The problem can be solved as follows:

$$\lim_{x \to 2^-} \frac{x^3 - 8}{|x - 2|} + 5x = \lim_{x \to 2^-} \left(\frac{(x - 2)(x^2 + 2x + 4)}{-(x - 2)} + 5x \right) = \lim_{x \to 2^-} \left(-x^2 - 2x - 4 + 5x \right)$$

$$\lim_{x \to 2^-} \left(-x^2 + 3x - 4 \right) = -4 + 6 - 4 = -2$$

Choice (2) is the answer.

6.34. The problem can be solved as follows:

$$\lim_{x \to \left(\frac{\pi}{2} \right)} \frac{\sin(x) + \cos(x)}{\cos(x)} = \frac{1^- + 0^-}{0^-} = \frac{1^-}{0^-} = -\infty$$

Choice (2) is the answer.

6.35. The problem can be solved as follows:

$$\lim_{x \to 0} \mathrm{arc}(\sin(x)) \equiv x$$

$$\lim_{x \to 0} 3x^4 + 2x^3 \equiv 2x^3$$

$$\lim_{x \to 0} \frac{3x^4 + 2x^3}{(\mathrm{arc}(\sin(x)))^3} \equiv \lim_{x \to 0} \frac{2x^3}{x^3} = \lim_{x \to 0} 2 = 2$$

Choice (2) is the answer.

6.36. The problem can be solved as follows:

$$\lim_{x \to -\infty} \left[\frac{2}{x + 1} \right] x = \lim_{x \to -\infty} [0^-] x = (-1)(-\infty) = +\infty$$

Choice (1) is the answer.

6.37. The problem can be solved as follows:

$$\lim_{x \to 3} \frac{x - 4}{\sqrt{x^2 - 4x + 3}} = \lim_{x \to 3^+} \frac{x - 4}{\sqrt{(x - 3)(x - 1)}} = \frac{-1}{0^+} = -\infty$$

Choice (2) is the answer.

6.38. From calculus, we know that:

$$\lim_{x \to \pm\infty} \sqrt{x^2 + ax + b} \equiv \left| x + \frac{a}{2} \right|$$

The problem can be solved as follows:

$$\lim_{x \to -\infty} \left(x + \sqrt{x^2 + 4x - 10} \right) \equiv \lim_{x \to -\infty} (x + |x + 2|) = \lim_{x \to -\infty} (x - x - 2) = \lim_{x \to -\infty} (-2) = -2$$

Choice (2) is the answer.

6.39. The problem can be solved as follows:

$$\lim_{x \to 2} \frac{4 - x^2}{6 - 2\sqrt{x^2 + 5}} = \frac{0}{0}$$

$$\xRightarrow{H} \lim_{x \to 2} \frac{\frac{d}{dx}\left(4 - x^2\right)}{\frac{d}{dx}\left(6 - 2\sqrt{x^2 + 5}\right)} = \lim_{x \to 2} \frac{-2x}{-2 \times \frac{2x}{2\sqrt{x^2+5}}} = \lim_{x \to 2} \sqrt{x^2 + 5} = 3$$

Choice (3) is the answer.

6.40. The problem can be solved as follows:

$$\lim_{x \to 0} \frac{\sin(2x)}{\sqrt{x + 1} - 1} = \frac{0}{0}$$

$$\xRightarrow{H} \lim_{x \to 0} \frac{\frac{d}{dx}\left(\sin(2x)\right)}{\frac{d}{dx}\left(\sqrt{x + 1} - 1\right)} = \lim_{x \to 0} \frac{2\cos(2x)}{\frac{1}{2\sqrt{x+1}}} = \frac{2 \times 1}{\frac{1}{2}} = 4$$

Choice (2) is the answer.

6.41. From calculus, we know that:

$$\lim_{x \to -\infty} \sqrt{x^4 + 2x^2 + x} \equiv x^2$$

$$\lim_{x \to \pm\infty} \left(a_m x^m + a_{m-1} x^{m-1} + \ldots + a_2 x^2 + a_1 x + a_0\right) \equiv a_m x^m$$

or:

$$\lim_{x \to \pm\infty} \left(a_m x^m + a_n x^n\right) \equiv a_m x^m \quad if \ m > n$$

The problem can be solved as follows:

$$\lim_{x \to -\infty} \sqrt{x^4 + 2x^2 + x} - x^2 = \lim_{x \to -\infty} \left(\sqrt{x^4 + 2x^2 + x} - x^2\right) \times \frac{\left(\sqrt{x^4 + 2x^2 + x} + x^2\right)}{\left(\sqrt{x^4 + 2x^2 + x} + x^2\right)}$$

$$= \lim_{x \to -\infty} \frac{x^4 + 2x^2 + x - x^4}{\left(\sqrt{x^4 + 2x^2 + x} + x^2\right)} \equiv \lim_{x \to -\infty} \frac{2x^2 + x}{(x^2 + x^2)} = \lim_{x \to -\infty} \frac{2x^2}{2x^2} = \lim_{x \to -\infty} 1 = 1$$

Choice (1) is the answer.

6.42. From calculus, we know that the limit of a function at a specific point (x_0) exits if:

$$\lim_{x \to x_0^-} f(x) = \lim_{x \to x_0^+} f(x)$$

Therefore, we must have:

$$\lim_{x \to (-3)^-} \frac{\left|x^2 - 9\right|}{x + 3} = \lim_{x \to (-3)^+} \frac{\left|x^2 - 9\right|}{x + 3} \tag{1}$$

$$\lim_{x \to (-3)^-} \frac{|x^2 - 9|}{x + 3} = \lim_{x \to (-3)^-} \frac{x^2 - 9}{x + 3} = \lim_{x \to (-3)^-} (x - 3) = -6 \qquad (2)$$

$$\lim_{x \to (-3)^+} \frac{|x^2 - 9|}{x + 3} = \lim_{x \to (-3)^+} \frac{-(x^2 - 9)}{x + 3} = \lim_{x \to (-3)^+} -(x - 3) = 6 \qquad (3)$$

$$\xrightarrow{(1), (2), (3)} -6 \neq 6 \Rightarrow \lim_{x \to -3} \frac{|x^2 - 9|}{x + 3} = \text{Undefined}$$

Choice (4) is the answer.

6.43. The problem can be solved as follows:

$$\lim_{x \to \frac{1}{2}} \frac{\tan\left(\frac{\pi x}{2}\right) - 1}{\cos(\pi x)} = \frac{0}{0}$$

$$\xrightarrow{H} \lim_{x \to \frac{1}{2}} \frac{\frac{d}{dx}\left(\tan\left(\frac{\pi x}{2}\right) - 1\right)}{\frac{d}{dx}(\cos(\pi x))} = \lim_{x \to \frac{1}{2}} \frac{\frac{\pi}{2}\left(1 + \tan^2\left(\frac{\pi x}{2}\right)\right)}{-\pi \, \sin(\pi x)} = \frac{\frac{\pi}{2}(1 + 1)}{-\pi \times 1} = -1$$

Choice (2) is the answer.

6.44. The problem can be solved as follows:

$$\lim_{x \to +\infty} \left(\sqrt{x + 5} - \sqrt{x + 1}\right) = \lim_{x \to +\infty} \left(\sqrt{x + 5} - \sqrt{x + 1}\right) \times \frac{\left(\sqrt{x + 5} + \sqrt{x + 1}\right)}{\left(\sqrt{x + 5} + \sqrt{x + 1}\right)}$$

$$= \lim_{x \to +\infty} \frac{x + 5 - (x + 1)}{\left(\sqrt{x + 5} + \sqrt{x + 1}\right)} = \lim_{x \to +\infty} \frac{4}{\left(\sqrt{x + 5} + \sqrt{x + 1}\right)} = 0$$

Choice (3) is the answer.

6.45. From trigonometry, we know that:

$$1 + \cos(2x) = 2\cos^2(x)$$

$$\tan(x) = \frac{\sin(x)}{\cos(x)}$$

$$\sin(2x) = 2\sin(x)\cos(x)$$

The problem can be solved as follows:

$$\lim_{x \to \frac{\pi}{2}} \frac{\tan(2x)\cos(x)}{1 + \cos(2x)} = \lim_{x \to \frac{\pi}{2}} \frac{\sin(2x)\cos(x)}{\cos(2x) \times 2\cos^2(x)} = \lim_{x \to \frac{\pi}{2}} \frac{2\sin(x)\cos^2(x)}{\cos(2x) \times 2\cos^2(x)} = \lim_{x \to \frac{\pi}{2}} \frac{\sin(x)}{\cos(2x)} = \frac{1}{-1} = -1$$

Choice (3) is the answer.

6.46. From calculus and trigonometry, we know that:

$$1 - \cos^2(x) = \sin^2(x)$$

Moreover, from application of Taylor series in limit, we know that:

$$\lim_{x \to 0^-} \sin(x) \equiv x$$

$$\lim_{x \to 0^-} \tan(x) \equiv x$$

The problem can be solved as follows:

$$\lim_{x \to 0} \frac{\tan(2x)}{\sqrt{1 - \cos(x)}} = \lim_{x \to 0} \frac{\tan(2x)}{\sqrt{1 - \cos(x)}} \times \frac{\sqrt{1 + \cos(x)}}{\sqrt{1 + \cos(x)}} = \lim_{x \to 0} \frac{\tan(2x)\sqrt{1 + \cos(x)}}{\sqrt{1 - \cos^2(x)}}$$

$$= \lim_{x \to 0^-} \frac{\tan(2x) \times \sqrt{2}}{\sqrt{\sin^2(x)}} = \lim_{x \to 0^-} \frac{\tan(2x) \times \sqrt{2}}{|\sin(x)|} = \lim_{x \to 0^-} \frac{\tan(2x) \times \sqrt{2}}{-\sin(x)} = \lim_{x \to 0^-} \frac{2x \times \sqrt{2}}{-x}$$

$$= \lim_{x \to 0^-} \left(-2\sqrt{2}\right) = -2\sqrt{2}$$

Choice (1) is the answer.

6.47. The problem can be solved as follows:

$$\lim_{x \to -\infty} \left(\sqrt[3]{n + 1000} - \sqrt[3]{n - 20}\right)$$

$$= \lim_{x \to -\infty} \left(\sqrt[3]{n + 1000} - \sqrt[3]{n - 20}\right) \times \frac{\left(\sqrt[3]{(n + 1000)^2} + \sqrt[3]{(n + 1000)(n - 20)} + \sqrt[3]{(n - 20)^2}\right)}{\left(\sqrt[3]{(n + 1000)^2} + \sqrt[3]{(n + 1000)(n - 20)} + \sqrt[3]{(n - 20)^2}\right)}$$

$$= \lim_{x \to -\infty} \frac{n + 1000 - (n - 20)}{\sqrt[3]{(n + 1000)^2} + \sqrt[3]{(n + 1000)(n - 20)} + \sqrt[3]{(n - 20)^2}} = \frac{1020}{+\infty} = 0$$

Choice (2) is the answer.

6.48. From application of Taylor series in limit, we know that:

$$\lim_{x \to 0} \sin(x) \equiv x$$

In addition, from trigonometry, we know that:

$$1 + \cos(x) = 2\cos^2\left(\frac{x}{2}\right)$$

$$\sin(x) = 2\sin\left(\frac{x}{2}\right)\cos\left(\frac{x}{2}\right)$$

The problem can be solved as follows:

$$\lim_{x \to \pi^+} \frac{\sin(\pi \sin(x)) \sin\left(\frac{x}{2}\right)}{\sqrt{1 + \cos(x)}} \equiv \lim_{x \to \pi^+} \frac{\pi \sin(x) \sin\left(\frac{x}{2}\right)}{\sqrt{2\cos^2\left(\frac{x}{2}\right)}} = \lim_{x \to \pi^+} \frac{\pi \times 2 \sin\left(\frac{x}{2}\right) \cos\left(\frac{x}{2}\right) \sin\left(\frac{x}{2}\right)}{\sqrt{2}\left|\cos\left(\frac{x}{2}\right)\right|}$$

$$= \lim_{x \to \pi^+} \frac{\pi \times 2 \sin^2\left(\frac{x}{2}\right) \cos\left(\frac{x}{2}\right)}{\sqrt{2}\left(-\cos\left(\frac{x}{2}\right)\right)} = \lim_{x \to \pi^+} \left(-\pi\sqrt{2}\sin^2\left(\frac{x}{2}\right)\right) = -\pi\sqrt{2} \times 1 = -\pi\sqrt{2}$$

Choice (1) is the answer.

6.49. From application of Taylor series in limit, we know that:

$$\lim_{\alpha \to 0} \sqrt[n]{1 + \alpha} \equiv \lim_{\alpha \to 0} 1 + \frac{\alpha}{n}$$

$$\lim_{x \to 0} \left(a_m x^m + a_{m-1} x^{m-1} + \dots + a_{m-n} x^{m-n} + a_{m-n-1} x^{m-n-1}\right) \equiv a_{m-n-1} x^{m-n-1}$$

or:

$$\lim_{x \to 0} \left(a_m x^m + a_n x^n\right) \equiv a_n x^n \quad if \ m > n$$

The problem can be solved as follows:

$$\lim_{x \to 0} \frac{\sqrt{1 + x^2} - \sqrt[4]{1 - 2x}}{2x^2 + 2x} \equiv \lim_{x \to 0} \frac{1 + \frac{x^2}{3} - \left(1 - \frac{2x}{4}\right)}{2x^2 + 2x} = \lim_{x \to 0} \frac{\frac{x^2}{3} + \frac{x}{2}}{2x^2 + 2x} \equiv \lim_{x \to 0} \frac{\frac{x}{2}}{2x} = \frac{1}{4}$$

Choice (1) is the answer

6.50. From trigonometry, we know that:

$$\sin^2(x) + \cos^2(x) = 1$$

The problem can be solved as follows:

$$\lim_{x \to 0} \frac{\sin^2(x) + \sin(x) + \cos^2(x) - \cos(x)}{\sin^2(x) - \sin(x) + \cos^2(x) - \cos(x)} = \lim_{x \to 0} \frac{1 + \sin(x) - \cos(x)}{1 - \sin(x) - \cos(x)} = \frac{0}{0}$$

$$\xrightarrow{H} \lim_{x \to 0} \frac{\frac{d}{dx}(1 + \sin(x) - \cos(x))}{\frac{d}{dx}(1 - \sin(x) - \cos(x))} = \lim_{x \to 0} \frac{\cos(x) + \sin(x)}{-\cos(x) + \sin(x)} = \frac{1 + 0}{-1 + 0} = -1$$

Choice (2) is the answer.

6.51. From application of Taylor series in limit, we know that:

$$\lim_{u(x) \to 0} \sin(u(x)) \equiv u(x)$$

The problem can be solved as follows:

$$\lim_{x \to 0} \frac{\cos(mx) - \cos(nx)}{x^2} = \frac{0}{0}$$

$$\xrightarrow{H} \lim_{x \to 0} \frac{\frac{d}{dx}(\cos(mx) - \cos(nx))}{\frac{d}{dx}(x^2)} = \lim_{x \to 0} \frac{-m \sin(mx) + n \sin(nx)}{2x} \equiv \lim_{x \to 0} \frac{-m(mx) + n(nx)}{2x}$$

$$= \lim_{x \to 0} \frac{-m^2 + n^2}{2} = \frac{n^2 - m^2}{2}$$

Choice (3) is the answer.

6.52. From application of Taylor series in limit, we know that:

$$\lim_{x \to 0} \sin(x) \equiv x$$

$$\lim_{x \to 0} \tan(x) \equiv x$$

The problem can be solved as follows:

$$\lim_{x \to 0} \frac{\sin(x) - x}{\tan(x) - x} = \frac{0}{0}$$

$$\xrightarrow{H} \lim_{x \to 0} \frac{\frac{d}{dx}(\sin(x) - x)}{\frac{d}{dx}(\tan(x) - x)} = \lim_{x \to 0} \frac{\cos(x) - 1}{1 + \tan^2(x) - 1} = \lim_{x \to 0} \frac{\cos(x) - 1}{\tan^2(x)} = \frac{0}{0}$$

$$\xrightarrow{H} \lim_{x \to 0} \frac{\frac{d}{dx}(\cos(x) - 1)}{\frac{d}{dx}(\tan^2(x))} = \lim_{x \to 0} \frac{-\sin(x)}{2\tan(x)(1 + \tan^2(x))} \equiv \lim_{x \to 0} \frac{-x}{2x(1 + x^2)} = \lim_{x \to 0} \frac{-1}{2(1 + x^2)}$$

$$= \frac{-1}{2(1 + 0)} = \frac{-1}{2}$$

Choice (1) is the answer.

6.53. From trigonometry, we know that:

$$1 + \cos^3(x) = (1 + \cos(x))(1 - \cos(x) + \cos^2(x))$$

$$1 - \cos^2(x) = (1 + \cos(x))(1 - \cos(x))$$

The problem can be solved as follows:

$$\lim_{x \to \pi} \frac{1 + \cos^3(x)}{1 - \cos^2(x)} = \frac{0}{0}$$

$$\Rightarrow \lim_{x \to \pi} \frac{1 + \cos^3(x)}{1 - \cos^2(x)} = \lim_{x \to \pi} \frac{(1 + \cos(x))(1 - \cos(x) + \cos^2(x))}{(1 + \cos(x))(1 - \cos(x))} = \lim_{x \to \pi} \frac{1 - \cos(x) + \cos^2(x)}{1 - \cos(x)}$$

$$= \frac{1 - (-1) + (-1)^2}{1 - (-1)} = \frac{3}{2}$$

Choice (1) is the answer.

6.54. Based on the information given in the problem, we have:

$$f(0) = 0$$

$$f(x) = x(-1)^{\left[\frac{1}{x}\right]}, \quad x \in \mathbb{R} - \{0\}$$

A function is right continuous at this given point of x_0 if:

$$\lim_{x \to x_0^+} f(x) = f(x_0)$$

Moreover, a function is left continuous at this given point of x_0 if:

$$\lim_{x \to x_0^-} f(x) = f(x_0)$$

In addition, a function is continuous at this given point of x_0 if:

$$\lim_{x \to x_0^+} f(x) = \lim_{x \to x_0^-} f(x) = f(x_0)$$

For the given function, we have:

$$\begin{cases} \lim_{x \to 0^-} f(x) = \lim_{x \to 0^-} x(-1)^{\left[\frac{1}{x}\right]} = 0 \times (\text{finite quantity}) = 0 \\ \lim_{x \to 0^+} f(x) = \lim_{x \to 0^+} x(-1)^{\left[\frac{1}{x}\right]} = 0 \times (\text{finite quantity}) = 0 \\ f(0) = 0 \end{cases}$$

$$\Rightarrow \lim_{x \to 0^+} f(x) = \lim_{x \to 0^-} f(x) = f(0)$$

Thus, the function is continuous at $x = 0$.

Choice (3) is the answer.

6.55. As we know from calculus:

$$\text{If } a > 1, k \in N \Rightarrow \lim_{n \to +\infty} \frac{n^k}{a^n} = 0$$

Hence:

$$\lim_{n \to +\infty} \frac{3n^2}{\sqrt{5^n}} = \lim_{n \to +\infty} 3 \frac{n^2}{(\sqrt{5})^n} = 3 \times 0 = 0$$

Choice (1) is the answer.

6.56. From trigonometry, we know that:

$$1 - \cos(x) = 2\sin^2\left(\frac{x}{2}\right)$$

Moreover, from application of Taylor series in limit, we know that:

$$\lim_{x \to 0} \sin^n(x) \equiv x^n$$

Thus:

$$\lim_{x \to 0} \frac{x^3 - \sin(x)(1 - \cos(x))}{x^3} = \lim_{x \to 0} \frac{x^3 - \sin(x)\left(2\sin^2\left(\frac{x}{2}\right)\right)}{x^3} \equiv \lim_{x \to 0} \frac{x^3 - x \times 2\left(\frac{x}{2}\right)^2}{x^3}$$

$$= \lim_{x \to 0} \frac{x^3 - \frac{x^3}{2}}{x^3} = \lim_{x \to 0} \frac{\frac{x^3}{2}}{x^3} = \lim_{x \to 0} \frac{1}{2} = \frac{1}{2}$$

Choice (2) is the answer.

6.57. From trigonometry and calculus, we know that:

$$\mathrm{arc}(\cos(1^-)) = 0^+$$

$$\frac{d}{dx}(\mathrm{arc}(\cos x)) = \frac{-1}{\sqrt{1 - x^2}}$$

$$\frac{d}{dx}\left(\sqrt{1 - x}\right) = \frac{-1}{2\sqrt{1 - x}}$$

The problem can be solved as follows:

$$\lim_{x \to 1} \frac{\mathrm{arc}(\cos x)}{\sqrt{1 - x}} = \frac{0^+}{0^+}$$

$$\overset{H}{\Longrightarrow} \lim_{x \to 1^-} \frac{\frac{d}{dx}(\mathrm{arc}(\cos x))}{\frac{d}{dx}\left(\sqrt{1 - x}\right)} = \lim_{x \to 1^-} \frac{\frac{-1}{\sqrt{1 - x^2}}}{\frac{-1}{2\sqrt{1 - x}}} = \lim_{x \to 1^-} \frac{\frac{1}{\sqrt{(1 - x)(1 + x)}}}{\frac{1}{2\sqrt{1 - x}}} = \lim_{x \to 1^-} \frac{2}{\sqrt{1 + x}} = \sqrt{2}$$

Choice (1) is the answer.

6.58. Based on the information given in the problem, we have:

$$\lim_{x \to 1} (x^2 - 1)\cot(x^n - 1) = \frac{1}{2} \tag{1}$$

From calculus, we know that:

$$\cot(x) = \frac{1}{\tan(x)}$$

From application of Taylor series in limit, we know that:

$$\lim_{u(x) \to 0} \tan(u(x)) \equiv u(x)$$

The problem can be solved as follows:

$$\lim_{x \to 1} \left(x^2 - 1\right) \cot(x^n - 1) = \lim_{x \to 1} \frac{x^2 - 1}{\tan(x^n - 1)} \equiv \lim_{x \to 1} \frac{x^2 - 1}{x^n - 1} = \frac{0}{0}$$

$$\xRightarrow{H} \lim_{x \to 1} \frac{2x}{nx^{n-1}} = \frac{2}{n} \qquad\qquad (2)$$

Solving (1) and (2):

$$\frac{2}{n} = \frac{1}{2} \Rightarrow n = 4$$

Choice (2) is the answer.

6.59. From trigonometry, we know that:

$$\sin(4x) = 2\sin(2x)\cos(2x)$$

$$\cot(x) = \frac{\cos(x)}{\sin(x)}$$

$$\sin(x - y) = \sin(x)\cos(y) - \cos(x)\sin(y)$$

The problem can be solved as follows:

$$\lim_{x \to 0} \sin(4x)(\cot(2x) - \cot(x)) = \lim_{x \to 0} 2\sin(2x)\cos(2x)\left(\frac{\cos(2x)}{\sin(2x)} - \frac{\cos(x)}{\sin(x)}\right)$$

$$= \lim_{x \to 0} 2\sin(2x)\cos(2x)\left(\frac{\sin(x)\cos(2x) - \cos(x)\sin(2x)}{\sin(2x)\sin(x)}\right)$$

$$= \lim_{x \to 0} 2\sin(2x)\cos(2x)\left(\frac{\sin(x - 2x)}{\sin(2x)\sin(x)}\right) = \lim_{x \to 0}(-2\cos(2x)) = -2$$

Choice (3) is the answer.

6.60. From application of Taylor series in limit, we know that:

$$\lim_{x \to 0} \sin(x) \equiv x - \frac{x^3}{6}$$

$$\lim_{x \to 0} \cos(x) \equiv 1 - \frac{x^2}{2}$$

The problem can be solved as follows:

$$\lim_{x \to 0} \frac{\sin(x) - x}{\frac{1}{2}\sin(2x) - x\cos(x)} \equiv \lim_{x \to 0} \frac{-\frac{x^3}{6}}{\frac{1}{2}\left(2x - \frac{(2x)^3}{6}\right) - x\left(1 - \frac{x^2}{2}\right)} = \lim_{x \to 0} \frac{-\frac{x^3}{6}}{-\frac{x^3}{6}} = \lim_{x \to 0} 1 = 1$$

Choice (3) is the answer.

6.61. The problem can be solved as follows:

$$\lim_{x \to \frac{\pi}{4}} \frac{1 - \sqrt[3]{\tan(x)}}{1 - 2\sin^2(x)} = \frac{0}{0}$$

$$\xrightarrow{H} \lim_{x \to \frac{\pi}{4}} \frac{\frac{d}{dx}\left(1 - \sqrt[3]{\tan(x)}\right)}{\frac{d}{dx}\left(1 - 2\sin^2(x)\right)} = \lim_{x \to \frac{\pi}{4}} \frac{-\frac{1 + \tan^2(x)}{3\sqrt[3]{\tan^2(x)}}}{-4\sin(x)\cos(x)} = \frac{\frac{1+1}{3 \times 1}}{4 \times \frac{\sqrt{2}}{2} \times \frac{\sqrt{2}}{2}} = \frac{1}{3}$$

Choice (1) is the answer.

6.62. From calculus and trigonometry, we know that:

$$\sin^2(x) + \cos^2(x) = 1$$

$$1 - \cos^3(x) = (1 - \cos(x))\left(1 + \cos(x) + \cos^2(x)\right)$$

From application of Taylor series in limit, we know that:

$$\lim_{x \to 0} \sin(x) \equiv x$$

$$\lim_{u(x) \to 0} \tan(u(x)) \equiv u(x)$$

The problem can be solved as follows:

$$\lim_{x \to 0} \frac{1 - \cos^3(x)}{\sin(x)\tan(2x)} = \frac{0}{0}$$

$$\Rightarrow \lim_{x \to 0} \frac{1 - \cos^3(x)}{\sin(x)\tan(2x)} = \lim_{x \to 0} \frac{(1 - \cos(x))(1 + \cos(x) + \cos^2(x))}{\sin(x)\tan(2x)} \times \frac{(1 - \cos(x))}{(1 + \cos(x))}$$

$$= \lim_{x \to 0} \frac{(1 - \cos^2(x))(1 + \cos(x) + \cos^2(x))}{\sin(x)\tan(2x)(1 + \cos(x))} = \lim_{x \to 0} \frac{\sin^2(x) \times (1 + 1 + 1)}{\sin(x)\tan(2x) \times (1 + 1)}$$

$$\lim_{x \to 0} \frac{3\sin^2(x)}{2\sin(x)\tan(2x)} \equiv \lim_{x \to 0} \frac{3x^2}{2x \times 2x} = \lim_{x \to 0} \frac{3}{4} = \frac{3}{4}$$

Choice (3) is the answer.

6.63. From trigonometry, we know that:

$$1 - \cos(x) = 2\sin^2\left(\frac{x}{2}\right)$$

$$\sin^2(x) + \cos^2(x) = 1$$

From application of Taylor series in limit, we know that:

$$\lim_{u(x) \to 0^+} \sin^n(u(x)) \equiv \lim_{u(x) \to 0^+} (u(x))^n$$

The problem can be solved as follows:

$$\lim_{x \to 0} \frac{1 - \sqrt{\cos(x)}}{1 - \cos(\sqrt{x})} = \lim_{x \to 0^+} \frac{1 - \sqrt{\cos(x)}}{1 - \cos(\sqrt{x})} \times \frac{1 + \sqrt{\cos(x)}}{1 + \sqrt{\cos(x)}} \times \frac{1 + \cos(\sqrt{x})}{1 + \cos(\sqrt{x})}$$

$$= \lim_{x \to 0^+} \frac{(1 - \cos(x))(1 + \cos(\sqrt{x}))}{(1 - \cos^2(\sqrt{x}))(1 + \sqrt{\cos(x)})} = \lim_{x \to 0^+} \frac{(1 - \cos(x))(1 + 1)}{(1 - \cos^2(\sqrt{x}))(1 + 1)}$$

$$= \lim_{x \to 0^+} \frac{1 - \cos(x)}{1 - \cos^2(\sqrt{x})} = \lim_{x \to 0^+} \frac{2 \sin^2\left(\frac{x}{2}\right)}{\sin^2(\sqrt{x})} \equiv \lim_{x \to 0^+} \frac{2\left(\frac{x}{2}\right)^2}{(\sqrt{x})^2} = \lim_{x \to 0^+} \frac{x}{2} = 0$$

Choice (1) is the answer.

References

1. Rahmani-Andebili, M. (2021). Calculus – Practice Problems, Methods, and Solutions, Springer Nature, 2021.
2. Rahmani-Andebili, M. (2021). Precalculus – Practice Problems, Methods, and Solutions, Springer Nature, 2021.

Abstract

In this chapter, the basic and advanced problems of derivatives and their applications are presented. The subjects include the definition of derivative, differentiation formulas, product rule, quotient rule, chain rule, derivatives of trigonometric functions, derivatives of exponential functions, derivatives of logarithm functions, derivatives of inverse trigonometric functions, derivatives of hyperbolic functions, implicit differentiation, higher-order derivatives, logarithmic differentiation, applications of derivatives, rates of change, critical points, minimum and maximum values, and absolute extrema. To help students study the chapter in the most efficient way, the problems are categorized in different levels based on their difficulty levels (easy, normal, and hard) and calculation amounts (small, normal, and large). Moreover, the problems are ordered from the easiest problem with the smallest computations to the most difficult problems with the largest calculations.

7.1. The current population of a specific animal in a jungle is about 820. How long will it take for the population to be 3280 if the growth constant is about 0.2 [1, 2]?

Difficulty level ● Easy ○ Normal ○ Hard
Calculation amount ● Small ○ Normal ○ Large
1) $4 \ln 2$
2) $10 \ln 2$
3) $2 \ln 2$
4) $2 \ln 10$

7.2. Which one of the following choices presents the nondifferentiable point(s) of the function below?

$$f(x) = \left| x(x+2)^2 (x-3)^3 \right|$$

Difficulty level ● Easy ○ Normal ○ Hard
Calculation amount ● Small ○ Normal ○ Large
1) $x = -2$
2) $x = 3$
3) $x = 0$
4) $x = -2, 0, 3$

7.3. Calculate the value of $f'(x = 1)$ if $f(x) = xe^x - e^x$.

Difficulty level ● Easy ○ Normal ○ Hard
Calculation amount ● Small ○ Normal ○ Large
1) 1
2) 0
3) $-e$
4) e

7.4. If $f(x) + g(x^3) = 5x - 1$ and $f'(1) = 2$, calculate the value of $g'(1)$.

Difficulty level   ● Easy   ○ Normal   ○ Hard
Calculation amount ● Small   ○ Normal   ○ Large
1) 1
2) −1
3) 2
4) −2

7.5. Determine the range of x where the function of $y(x) = 1 - 4x^2$ is ascending.

Difficulty level   ● Easy   ○ Normal   ○ Hard
Calculation amount ● Small   ○ Normal   ○ Large
1) $x < 0$
2) $x > 0$
3) $-2 < x < 2$
4) $-4 < x < 4$

7.6. Determine the derivative of the function below at $x = \dfrac{1}{4}$.

$$f(x) = \frac{x - \sqrt{x}}{1 - \sqrt{x}}$$

Difficulty level   ● Easy   ○ Normal   ○ Hard
Calculation amount ● Small   ○ Normal   ○ Large
1) −1
2) $-\dfrac{1}{2}$
3) $\dfrac{1}{2}$
4) 1

7.7. Determine the first derivative of the function of $(x^{100} + x^{50} + 50x^2 + 50x + 1)^{10}$ at $x = 0$.

Difficulty level   ● Easy   ○ Normal   ○ Hard
Calculation amount ○ Small   ● Normal   ○ Large
1) 100
2) 200
3) 400
4) 500

7.8. Calculate the derivative of the function of $f(x) = \tan^3(2x)$ at $\dfrac{\pi}{12}$.

Difficulty level   ● Easy   ○ Normal   ○ Hard
Calculation amount ○ Small   ● Normal   ○ Large
1) $\dfrac{4}{3}$
2) $\dfrac{4}{9}$
3) $\dfrac{8}{3}$
4) $\dfrac{8}{9}$

7.9. What is the first derivative of the inverse function of $f(x) = x^3 + x - 2$ at a point with the length of zero on the inverse function?

Difficulty level   ○ Easy   ● Normal   ○ Hard
Calculation amount   ● Small   ○ Normal   ○ Large

1) -1
2) 0
3) 1
4) $\dfrac{1}{4}$

7.10. If $f(x) = x^5 + 3x^3 + x + 1$, then calculate the first derivative of the inverse function at a point with the length of six on the inverse function.

Difficulty level   ○ Easy   ● Normal   ○ Hard
Calculation amount   ● Small   ○ Normal   ○ Large

1) $\dfrac{1}{5}$
2) $\dfrac{1}{15}$
3) $\dfrac{1}{6}$
4) $\dfrac{1}{16}$

7.11. For the following function, calculate the value of $(f^{-1}(x))'$ for $x = 2$.

$$f(x) = \frac{4x^3}{x^2 + 1}$$

Difficulty level   ○ Easy   ● Normal   ○ Hard
Calculation amount   ● Small   ○ Normal   ○ Large

1) $\dfrac{112}{25}$
2) $\dfrac{25}{112}$
3) $\dfrac{1}{4}$
4) 4

7.12. Calculate the value of the limit below if $f(x) = x \tan x$.

$$\lim_{x \to \frac{\pi}{4}} \frac{f(x) - f\left(\frac{\pi}{4}\right)}{x - \frac{\pi}{4}}$$

Difficulty level   ○ Easy   ● Normal   ○ Hard
Calculation amount   ● Small   ○ Normal   ○ Large

1) 1
2) $1 - \dfrac{\pi}{4}$
3) $1 + \dfrac{\pi}{4}$
4) $1 + \dfrac{\pi}{2}$

7.13. If the function of $f(x) = |x^3 - 3x + a|$ does not have a derivate at $x = 2$, calculate the value of a.

Difficulty level   ○ Easy   ● Normal   ○ Hard

Calculation amount   ● Small   ○ Normal   ○ Large

1) 2

2) −2

3) 1

4) −1

7.14. Calculate the value of $f'(2) + f'(4)$ if $f(x) = |x^2 - 6|$.

Difficulty level   ○ Easy   ● Normal   ○ Hard

Calculation amount   ● Small   ○ Normal   ○ Large

1) −8

2) 8

3) −4

4) 4

7.15. If $f'(x) = \frac{5}{x}$, calculate the first derivative of $f(x^5)$.

Difficulty level   ○ Easy   ● Normal   ○ Hard

Calculation amount   ● Small   ○ Normal   ○ Large

1) $-\dfrac{5}{x}$

2) $-\dfrac{25}{x}$

3) $\dfrac{25}{x}$

4) $\dfrac{5}{x^5}$

7.16. If the first derivative of $f(\sin(x))$ is equal to $\cos^3(x)$, determine the value of $f'(x)$.

Difficulty level   ○ Easy   ● Normal   ○ Hard

Calculation amount   ● Small   ○ Normal   ○ Large

1) $1 + x^2$

2) $1 - x^2$

3) x^3

4) $-x^3$

7.17. Calculate the derivative of the function of $f(x) = \text{arc}(\tan(3x))$ at $x = \dfrac{1}{3}$.

Difficulty level   ○ Easy   ● Normal   ○ Hard

Calculation amount   ● Small   ○ Normal   ○ Large

1) $\dfrac{3}{2}$

2) $\dfrac{4}{3}$

3) $\dfrac{2}{3}$

4) $\dfrac{3}{4}$

7.18. If $f\left(\dfrac{1}{t}\right) + g(\sqrt{t}) = t^2 + 1$ and $g'(1) = 5$, calculate the value of $f'(1)$.

Difficulty level   ○ Easy   ● Normal   ○ Hard

Calculation amount   ○ Small   ● Normal   ○ Large

1) 1

2) 2

3) $\dfrac{1}{2}$

4) $-\dfrac{1}{2}$

7.19. If $2\cos(y) - \sin(x+y) + 2 = 0$, calculate the value of y'_x at $(0, \pi)$.

Difficulty level   ○ Easy   ● Normal   ○ Hard

Calculation amount   ○ Small   ● Normal   ○ Large

1) $\dfrac{1}{2}$

2) $-\dfrac{1}{2}$

3) -1

4) 1

7.20. The equation of a curve is given by $x^3 + y^3 = 16$. Calculate the second derivate of y with respect to x.

Difficulty level   ○ Easy   ● Normal   ○ Hard

Calculation amount   ○ Small   ● Normal   ○ Large

1) $-\dfrac{16y}{x^5}$

2) $\dfrac{16x}{y^5}$

3) $-\dfrac{32y}{x^5}$

4) $-\dfrac{32x}{y^5}$

7.21. If $x = 2 + 3\sin(t)$ and $y - 3 - 2\cos(t)$, calculate the value of y'_x for $t - \dfrac{\pi}{6}$.

Difficulty level   ○ Easy   ● Normal   ○ Hard

Calculation amount   ○ Small   ● Normal   ○ Large

1) $\dfrac{2\sqrt{3}}{9}$

2) $\dfrac{2\sqrt{3}}{3}$

3) $\dfrac{2\sqrt{2}}{3}$

4) $\dfrac{4\sqrt{2}}{3}$

7.22. If $x = t^2 + t$ and $y = t^2 - 2t$, calculate the value of $x'_y + y'_x$ for $t = -1$.

Difficulty level   ○ Easy   ● Normal   ○ Hard

Calculation amount   ○ Small   ● Normal   ○ Large

1) $\dfrac{11}{4}$

2) $\dfrac{13}{4}$

3) $\dfrac{15}{4}$

4) $\dfrac{17}{4}$

7.23. Calculate the value of $f'(4)$ if we know that:

$$\lim_{h \to 0} \frac{f(x+h) - f(x-h)}{h} = 2\sqrt{x}$$

Difficulty level   ○ Easy   ● Normal   ○ Hard
Calculation amount   ○ Small   ● Normal   ○ Large

1) $\dfrac{2}{3}$

2) $\dfrac{4}{3}$

3) 4

4) 2

7.24. Which one of the choices is true about the function of $f(x) = x^2 |x|$ at $x = 0$?
Difficulty level   ○ Easy   ● Normal   ○ Hard
Calculation amount   ○ Small   ● Normal   ○ Large
1) The first derivative exists, but the second derivative does not.
2) The second derivative exists, but the first derivative does not.
3) The first and second derivatives do not exist.
4) The first and second derivatives exist.

7.25. The function below is differentiable at $x = \dfrac{\pi}{4}$. Determine the value of b.

$$f(x) = \begin{cases} \sin^2(x) - \cos(2x) & 0 < x \leq \dfrac{\pi}{4} \\ a\tan(x) + b\sin(2x) & \dfrac{\pi}{4} < x < \dfrac{\pi}{2} \end{cases}$$

Difficulty level   ○ Easy   ● Normal   ○ Hard
Calculation amount   ○ Small   ● Normal   ○ Large
1) -1

2) $-\dfrac{1}{2}$

3) $\dfrac{1}{2}$

4) 1

7.26. The function below is differentiable everywhere on $\mathbb{R}$ domain. Determine the value of b.

$$f(x) = \begin{cases} ax + b & x < -1 \\ x^2 + a & x \geq -1 \end{cases}$$

Difficulty level   ○ Easy   ● Normal   ○ Hard
Calculation amount   ○ Small   ● Normal   ○ Large
1) 2
2) 1
3) -2
4) -3

7.27. Calculate the derivative of the function below.

$$f(x) = \frac{(2x-1)^2}{2x^2}$$

Difficulty level   ○ Easy   ● Normal   ○ Hard
Calculation amount   ○ Small ● Normal   ○ Large

1) $\dfrac{2x-1}{2x^3}$

2) $\dfrac{2x-1}{x^3}$

3) $\dfrac{2x+1}{x^3}$

4) $\dfrac{2x+1}{2x^3}$

7.28. Calculate the derivative of the function below.

$$f(x) = \frac{\sin(x)}{1 + \tan^2(x)}$$

Difficulty level   ○ Easy   ● Normal   ○ Hard
Calculation amount   ○ Small ● Normal   ○ Large

1) $\dfrac{5}{4}$

2) $-\dfrac{5}{4}$

3) $\dfrac{5}{8}$

4) $-\dfrac{5}{8}$

7.29. For the function below, calculate the value of $f'(x = 2)$.

$$f(x) = \left(x^2 - 5x + 6\right)\mathrm{arc}\left(\sin\left(\frac{1}{x}\right)\right)$$

Difficulty level   ○ Easy   ● Normal   ○ Hard
Calculation amount   ○ Small ● Normal   ○ Large

1) $-\dfrac{\pi}{3}$

2) $\dfrac{2\pi}{3}$

3) $-\dfrac{\pi}{6}$

4) $\dfrac{\pi}{4}$

7.30. For the following function, calculate the value of $f'(x = -3)$.

$$f(x) = \left(x^2 + 2x - 3\right)\frac{g(x+2)}{(x^3+1)g(2x+5)}$$

Difficulty level   ○ Easy   ● Normal   ○ Hard
Calculation amount   ○ Small ● Normal   ○ Large

1) $-\dfrac{13}{2}$

2) $\dfrac{13}{2}$

3) $-\dfrac{2}{13}$

4) $\dfrac{2}{13}$

7.31. For what value of m, the line of $y = 2x + 1$ is tangent to a curve with the following function:

$$y = \frac{-1 + x^2}{m + x}$$

Difficulty level   ○ Easy   ● Normal   ○ Hard
Calculation amount   ○ Small   ● Normal   ○ Large

1) $\dfrac{3}{4}$

2) $\pm\dfrac{\sqrt{3}}{8}$

3) $\dfrac{1}{9}$

4) $\pm\dfrac{\sqrt{3}}{2}$

7.32. Determine the third derivate of $f(x) = x^4 \, |x|$.
Difficulty level   ○ Easy   ● Normal   ○ Hard
Calculation amount   ○ Small   ● Normal   ○ Large
1) $-60x^2$
2) $60x^2$
3) $-60x|x|$
4) $60x|x|$

7.33. Determine the value of the parameter of "a" if the derivative of $\sqrt{x + a}$ for $x = 2$ is $\dfrac{1}{4}$.
Difficulty level   ○ Easy   ● Normal   ○ Hard
Calculation amount   ○ Small   ● Normal   ○ Large
1) -2
2) -1
3) 1
4) 2

7.34. Calculate the derivative of $y = \ln e^{\sqrt{\sin(x)}}$ at $x = \dfrac{\pi}{6}$.
Difficulty level   ○ Easy   ● Normal   ○ Hard
Calculation amount   ○ Small   ● Normal   ○ Large

1) $\dfrac{\sqrt{3}}{8}$

2) $\dfrac{\sqrt{6}}{8}$

3) $\dfrac{\sqrt{3}}{4}$

4) $\dfrac{\sqrt{6}}{4}$

7.35. Determine the maximum value of the function of $y(x) = x^3 - 3x^2 - 9x + 5$ in the range of $[-2, 2]$.

Difficulty level   ○ Easy   ● Normal   ○ Hard
Calculation amount   ○ Small   ● Normal   ○ Large
1) 9
2) 10
3) 12
4) 17

7.36. Which one of the choices is correct about the function below in its one period?

$$y(x) = \frac{1 - \sin(x)}{\cos(x)}$$

Difficulty level   ○ Easy   ● Normal   ○ Hard
Calculation amount   ○ Small   ● Normal   ○ Large
1) The function is always ascending.
2) The function is always descending.
3) The function has one minimum point.
4) The function has one maximum point.

7.37. Determine the value of $f'(x)g(x) - f(x)g'(x)$ if we have the following functions:

$$f(x) = \left(\sqrt{1 + x^2} - x\right)^5, \quad g(x) = \frac{1}{\left(\sqrt{1 + x^2} + x\right)^5}$$

Difficulty level   ○ Easy   ● Normal   ○ Hard
Calculation amount   ○ Small   ● Normal   ○ Large
1) -1
2) 0
3) 1
4) 2

7.38. Calculate the first derivative of the following function for $x = \frac{4}{3}$:

$$y(x) = \frac{1}{\sqrt{1 + x^2}\left(x + \sqrt{1 + x^2}\right)}$$

Difficulty level   ○ Easy   ● Normal   ○ Hard
Calculation amount   ○ Small   ○ Normal   ● Large
1) $\frac{-27}{125}$
2) $\frac{-9}{25}$
3) $\frac{18}{25}$
4) $\frac{54}{125}$

7.39. On a curve with the function of $y = x^3 - 6x + 12$, two tangent lines, parallel to x-axis, have been drawn. Determine the distance between these two lines.

Difficulty level   ○ Easy   ○ Normal   ● Hard
Calculation amount   ○ Small   ● Normal   ○ Large
1) 14
2) 6
3) $4\sqrt{2}$
4) $8\sqrt{2}$

7.40. For the function below, calculate the value of $f'(x = -1)$.

$$f(x) = \begin{cases} \dfrac{(x+1)^5}{|x+1|} & x \neq -1 \\ 0 & x = -1 \end{cases}$$

Difficulty level   ○ Easy   ○ Normal   ● Hard
Calculation amount   ○ Small   ● Normal   ○ Large
1) 0
2) 1
3) -1
4) 5

7.41. The point $M(x, y)$ is moving on the curve of $y = \sqrt{x + 8}$. Determine the changing rate of the distance of the point from the origin when $x = 7$.

Difficulty level   ○ Easy   ○ Normal   ● Hard
Calculation amount   ○ Small   ● Normal   ○ Large
1) $\dfrac{15}{16}$
2) $\dfrac{15}{8}$
3) $\dfrac{3}{7}$
4) $\dfrac{5}{4}$

7.42. Determine the derivative of $f\left(\sqrt{|-x| + 3}\right)$ if we have the relation below.

$$\lim_{x \to 2} \frac{f(x) - f(2)}{x - 2} = -\frac{1}{3}$$

Difficulty level   ○ Easy   ○ Normal   ● Hard
Calculation amount   ○ Small   ● Normal   ○ Large
1) $\dfrac{1}{6}$
2) $\dfrac{1}{12}$
3) $-\dfrac{1}{6}$
4) $-\dfrac{1}{12}$

7.43. Determine the value of $f'(-1)$ if we have:

$$f(x) = \frac{(x+1)h(x)}{(2x+1)h(2x+1)}, \quad h(-1) \neq 0$$

Difficulty level   ○ Easy   ○ Normal   ● Hard
Calculation amount   ○ Small   ● Normal   ○ Large
1) -2
2) -1
3) 1
4) 2

7.44. Determine the value of the parameter of "a" so that the function of $f(x) = \cos^2(x) + \sqrt{3}\sin(x) + a$ has an extremum point with the width of $y = \frac{3}{4}$ in the range of $0 < x < \frac{\pi}{2}$.

Difficulty level   ○ Easy   ○ Normal   ● Hard
Calculation amount   ○ Small   ● Normal   ○ Large
1) 1
2) $\frac{1}{2}$
3) $-\frac{1}{2}$
4) -1

7.45. Calculate the first derivative of the function of $y(x) = x^x$ for $x = e$.
Difficulty level   ○ Easy   ○ Normal   ● Hard
Calculation amount   ○ Small   ● Normal   ○ Large
1) e
2) e^{e-1}
3) $2e^e$
4) e^e

7.46. Calculate the first derivative of the function of $y(x) = x^{\ln x}$.
Difficulty level   ○ Easy   ○ Normal   ● Hard
Calculation amount   ○ Small   ● Normal   ○ Large
1) $x^{\ln x} \ln x$
2) $2x^{\ln x} \ln x$
3) $\frac{2x^{\ln x} \ln x}{x}$
4) $(2x^{\ln x} \ln x)^{-x}$

7.47. Calculate the n-th derivative of the function below.

$$y(x) = \frac{1}{x}$$

Difficulty level   ○ Easy   ○ Normal   ● Hard
Calculation amount   ○ Small   ● Normal   ○ Large
1) 0
2) $\frac{n!}{x^{n+1}}$
3) $(-1)^n \frac{n!}{x^n}$
4) $(-1)^n \frac{n!}{x^{n+1}}$

7.48. Determine the equation of the line which is perpendicular on the curve of $y(x) = x^{2x}$ at $(1, 1)$.

1) $x + 2y - 3 = 0$
2) $2x + y - 3 = 0$
3) $x + y - 2 = 0$
4) $x - y = 0$

7.49. Determine the angle between the right and left tangent lines of the function below at $x = 1$.

$$f(x) = \begin{cases} x^3 & x > 1 \\ \sqrt{x} & x \leq 1 \end{cases}$$

1) $\dfrac{\pi}{4}$

2) $\dfrac{2\pi}{3}$

3) $\dfrac{\pi}{3}$

4) $\dfrac{3\pi}{4}$

References

1. Rahmani-Andebili, M. (2021). Calculus – Practice Problems, Methods, and Solutions, Springer Nature, 2021.
2. Rahmani-Andebili, M. (2021). Precalculus – Practice Problems, Methods, and Solutions, Springer Nature, 2021.

Abstract

In this chapter, the problems of the seventh chapter are fully solved, in detail, step-by-step, and with different methods.

8.1. Based on the information given in the problem, we have [1, 2]:

$$y(0) = 820$$

$$y(t) = 3280$$

$$k = 0.2$$

The population of the animal in the jungle at time t can be calculated as follows:

$$y(t) = y(0)e^{kt}$$

By putting the quantities in the formula, we have:

$$3280 = 820e^{0.2t}$$

$$\Rightarrow 4 = e^{0.2t} \Rightarrow \ln 4 = 0.2t \Rightarrow t = \frac{\ln 4}{0.2} = 5\ln 4$$

$$\Rightarrow t = 10\ln 2$$

Choice (2) is the answer.

In this problem, the rules below were used.

$$\ln e^b = b$$

$$\ln a^b = b\ln a$$

8.2. An absolute function is nondifferentiable at the simple roots of the equation inside the absolute notation.

As can be noticed from the equation inside the absolute notation, $x = 0$ is the only simple root. Thus, the absolute function is not differentiable at this point.

$$f(x) = \left| x(x+2)^2(x-3)^3 \right|$$

Choice (3) is the answer.

8.3. From list of derivative rules, we know that:

$$f(x) = e^x \Rightarrow f'(x) = e^x$$

$$f(x) = u(x)v(x) \Rightarrow f'(x) = u'(x)v(x) + u(x)v'(x)$$

Based on the information given in the problem, we have:

$$f(x) = xe^x - e^x$$

The problem can be solved as follows:

$$f'(x) = e^x + xe^x - e^x = xe^x$$

$$f'(1) = 1e^1 \Rightarrow f'(1) = e$$

Choice (4) is the answer.

8.4. From list of derivative rules, we know that:

$$h(x) = g(u(x)) \Rightarrow h'(x) = u'(x)g'(u(x))$$

Based on the information given in the problem, we have:

$$f'(1) = 2$$

$$f(x) + g(x^3) = 5x - 1$$

The problem can be solved as follows:

$$f(x) + g(x^3) = 5x - 1 \xrightarrow{\frac{d}{dx}} f'(x) + 3x^2 g'(x^3) = 5$$

$$\xrightarrow{x = 1} f'(1) + 3g'(1) = 5$$

$$\xrightarrow{f'(1) = 2} 2 + 3g'(1) = 5$$

$$\Rightarrow g'(1) = 1$$

Choice (1) is the answer.

8.5. From list of derivative rules, we know that:

$$f(x) = ax^n \Rightarrow f'(x) = anx^{n-1}$$

A function is ascending in a given range if its derivative is positive. Therefore:

$$y(x) = 1 - 4x^2 \Rightarrow y'(x) = -8x > 0 \Rightarrow x < 0$$

Choice (1) is the answer.

8.6. From list of derivative rules, we know that:

$$f(x) = \sqrt{x} \Rightarrow f'(x) = \frac{1}{2\sqrt{x}}$$

First, we should simplify the function as follows:

$$f(x) = \frac{x - \sqrt{x}}{1 - \sqrt{x}} = \frac{\sqrt{x}(\sqrt{x} - 1)}{1 - \sqrt{x}} = -\sqrt{x}$$

Therefore:

$$\Rightarrow f'(x) = -\frac{1}{2\sqrt{x}} \Rightarrow f'\left(\frac{1}{4}\right) = -\frac{1}{2\sqrt{\frac{1}{4}}} = -1$$

Choice (1) is the answer.

8.7. From list of derivative rules, we know that:

$$f(x) = u^n(x) \Rightarrow f'(x) = nu'(x)u^{n-1}(x)$$

Therefore:

$$f(x) = \left(x^{100} + x^{50} + 50x^2 + 50x + 1\right)^{10}$$

$$\Rightarrow f'(x) = 10\left(100x^{99} + 50x^{49} + 100x + 50\right)\left(x^{100} + x^{50} + 50x^2 + 50x + 1\right)^9$$

$$\Rightarrow f'(0) = 10(0 + 0 + 0 + 50)(0 + 0 + 0 + 0 + 1)^9 = 500$$

Choice (4) is the answer.

8.8. From list of derivative rules, we know that:

$$f(x) = \tan^n(u(x)) \Rightarrow f'(x) = nu'(x)\left(1 + \tan^2(u(x))\right)\tan^{n-1}(u(x))$$

Therefore:

$$f(x) = \tan^3(2x) \Rightarrow f'(x) = 3 \times 2\left(1 + \tan^2(2x)\right)\tan^2(2x)$$

$$\Rightarrow f'\left(\frac{\pi}{12}\right) = 3 \times 2\left(1 + \tan^2\left(\frac{\pi}{6}\right)\right)\tan^2\left(\frac{\pi}{6}\right) = 6 \times \left(1 + \frac{1}{3}\right) \times \frac{1}{3} = \frac{8}{3}$$

Choice (3) is the answer.

8.9. The derivative of the inverse function of $f(x)$ at a point with the length of "b" on the inverse function can be calculated as follows:

$$\left(f^{-1}\right)'(b) = \frac{1}{f'(a)}$$

On the other hand, we know that:

$$f(a) = b \Leftrightarrow f^{-1}(b) = a$$

Therefore, for the function of $f(x) = x^3 + x - 2$ and the point with the length of zero ($b = 0$) on the inverse function, we can calculate "a" as follows:

$$0 = a^3 + a - 2 \Rightarrow a = 1$$

Also, we have:

$$f(x) = x^3 + x - 2 \Rightarrow f'(x) = 3x^2 + 1$$

Hence:

$$\left(f^{-1}\right)'(0) = \frac{1}{f'(1)} = \frac{1}{3 \times 1^2 + 1} = \frac{1}{4}$$

Choice (4) is the answer.

8.10. As we know, the derivative of the inverse function of $f(x)$ at a point with the length of "b" on the inverse function can be calculated as follows:

$$\left(f^{-1}\right)'(b) = \frac{1}{f'(a)}$$

Moreover:

$$f(a) = b \Leftrightarrow f^{-1}(b) = a$$

Thus, for the function of $f(x) = x^5 + 3x^3 + x + 1$ and the point with the length of six ($b = 6$) on the inverse function, we can calculate "a" as follows:

$$6 = a^5 + 3a^3 + a + 1 \Rightarrow a = 1$$

On the other hand, we have:

$$f(x) = x^5 + 3x^3 + x + 1 \Rightarrow f'(x) = 5x^4 + 9x^2 + 1$$

Hence:

$$\left(f^{-1}\right)'(6) = \frac{1}{f'(1)} = \frac{1}{5 \times 1^4 + 9 \times 1^2 + 1} = \frac{1}{15}$$

Choice (2) is the answer.

8.11. As we know, the derivative of the inverse function of $f(x)$ at a point with the length of "b" on the inverse function can be calculated as follows:

$$\left(f^{-1}\right)'(b) = \frac{1}{f'(a)}$$

In addition, we know that:

$$f(a) = b \Leftrightarrow f^{-1}(b) = a$$

Hence, for the function of $f(x) = \frac{4x^3}{x^2+1}$ and the point with the length of two ($b = 2$) on the inverse function, we can calculate "a" as follows:

$$2 = \frac{4a^3}{a^2 + 1} \Rightarrow a = 1$$

On the other hand, we can write:

$$f(x) = \frac{4x^3}{x^2 + 1} \Rightarrow f'(x) = \frac{12x^2(x^2 + 1) - (2x)(4x^3)}{(x^2 + 1)^2}$$

Thus:

$$\left(f^{-1}\right)'(2) = \frac{1}{f'(1)} = \frac{\left(1^2 + 1\right)^2}{12 \times 1^2\left(1^2 + 1\right) - (2 \times 1)\left(4 \times 1^3\right)} = \frac{4}{24 - 8} = \frac{1}{4}$$

Choice (3) is the answer.

8.12. We know that:

$$\frac{d}{dx}(u(x)v(x)) = u'(x)v(x) + u(x)v'(x) \tag{1}$$

$$\frac{d}{dx}(\tan x) = 1 + \tan^2 x$$

$$\tan \frac{\pi}{4} = 1$$

The first derivative of a function is defined as follows:

$$\lim_{x \to x_0} \frac{f(x) - f(x_0)}{x - x_0}$$

Therefore, the value of the following limit for the function of $f(x) = x \tan x$ is equal to the first derivative of the function at $x_0 = \frac{\pi}{4}$.

$$\lim_{x \to \frac{\pi}{4}} \frac{f(x) - f\left(\frac{\pi}{4}\right)}{x - \frac{\pi}{4}} = f'(x_0)$$

The first derivative of the function can be calculated as follows:

$$f'(x) = \tan x + (1 + \tan^2 x)x$$

For $x_0 = \frac{\pi}{4}$, we have:

$$f'\left(\frac{\pi}{4}\right) = \tan\frac{\pi}{4} + \left(1 + \tan^2\frac{\pi}{4}\right)\frac{\pi}{4} = 1 + \frac{\pi}{2}$$

Choice (4) is the answer.

8.13. Based on the information given in the problem, we have:

$$f(x) = \left|x^3 - 3x + a\right|$$

A derivative of an absolute function does not exist at its simple roots. Therefore, we need to solve the equation below:

$$f(2) = 0$$

$$\Rightarrow 2^3 - 3 \times 2 + a = 0 \Rightarrow 8 - 6 + a = 0 \Rightarrow a = -2$$

Choice (2) is the answer.

8.14. From list of derivative rules, we know that:

$$f(x) = |u(x)| \Rightarrow f'(x) = \frac{u'(x)u(x)}{|u(x)|}$$

Based on the information given in the problem, we have:

$$f(x) = \left|x^2 - 6\right|$$

Therefore:

$$\Rightarrow f'(x) = \frac{2x \times (x^2 - 6)}{|x^2 - 6|}$$

$$\Rightarrow f'(2) = \frac{4 \times (4 - 6)}{|4 - 6|} = -4 \quad \text{and} \quad f'(4) = \frac{8 \times (16 - 6)}{|16 - 6|} = 8$$

$$\Rightarrow f'(2) + f'(4) = -4 + 8 = 4$$

Choice (4) is the answer.

8.15. From list of derivative rules, we know that:

$$f(x) = g(h(x)) \Rightarrow f'(x) = h'(x)g'(h(x))$$

Based on the information given in the problem, we have:

$$f'(x) = \frac{5}{x}$$

The problem can be solved as follows:

$$(f(x^5))' = 5x^4 \times f'(x^5) = 5x^4 \times \frac{5}{x^5} \Rightarrow (f(x^5))' = \frac{25}{x}$$

Choice (3) is the answer.

8.16. From list of derivative rules, we have:

$$(f(g(x)))' = g'(x)f'(g(x))$$

Based on the information given in the problem, we have:

$$(f(\sin(x)))' = \cos^3(x)$$

The problem can be solved as follows:

$$(f(\sin(x)))' = \cos^3(x) \Rightarrow \cos(x) \times f'(\sin(x)) = \cos^3(x) \Rightarrow f'(\sin(x)) = \cos^2(x)$$

$$\Rightarrow f'(\sin(x)) = 1 - \sin^2(x) \Rightarrow f'(x) = 1 - x^2$$

Choice (2) is the answer.

8.17. From list of derivative rules, we know that:

$$f(x) = \text{arc}(\tan(u(x))) \Rightarrow f'(x) = \frac{u'(x)}{1 + u^2(x)}$$

Therefore:

$$f(x) = \text{arc}(\tan(3x)) \Rightarrow f'(x) = \frac{3}{1 + 9x^2}$$

$$f'\left(\frac{1}{3}\right) = \frac{3}{1 + 9\left(\frac{1}{3}\right)^2} = \frac{3}{2}$$

Choice (1) is the answer.

8.18. From list of derivative rules, we know that:

$$f(x) = ax^n \Rightarrow f'(x) = anx^{n-1}$$

$$f(x) = g(h(x)) \Rightarrow f'(x) = h'(x)g'(h(x))$$

Based on the information given in the problem, we have:

$$f\left(\frac{1}{t}\right) + g(\sqrt{t}) = t^2 + 1$$

$$g'(1) = 5$$

The problem can be solved as follows:

$$f\left(\frac{1}{t}\right) + g\left(\sqrt{t}\right) = t^2 + 1 \xrightarrow{\frac{d}{dx}} \left(-\frac{1}{t^2}\right) \times f'\left(\frac{1}{t}\right) + \frac{1}{2\sqrt{t}} \times g'\left(\sqrt{t}\right) = 2t$$

$$t = 1 \Rightarrow -f'(1) + \frac{1}{2}g'(1) = 2 \xrightarrow{g'(1) = 5} -f'(1) + \frac{5}{2} = 2$$

$$\Rightarrow f'(1) = \frac{1}{2}$$

Choice (3) is the answer.

8.19. From derivative rules, we know that:

$$f(x, y) = 0 \Rightarrow y'_x = -\frac{f'_x(x, y)}{f'_y(x, y)} = -\frac{\frac{d}{dx}f(x, y)}{\frac{d}{dy}f(x, y)}$$

Based on the information given in the problem, we have:

$$2\cos(y) - \sin(x + y) + 2 = 0$$

The problem can be solved as follows:

$$y'_x = -\frac{\frac{d}{dx}(2\cos(y) - \sin(x + y) + 2)}{\frac{d}{dy}(2\cos(y) - \sin(x + y) + 2)} = -\frac{-\cos(x + y)}{-2\sin(y) - \cos(x + y)} = -\frac{\cos(x + y)}{2\sin(y) + \cos(x + y)}$$

$$\xrightarrow{(x, y) = (0, \pi)} y'_x = -\frac{-1}{0 - 1}$$

$$\Rightarrow y'_x = -1$$

Choice (3) is the answer.

8.20. From derivative rules, we know that:

$$f(x, y) = 0 \Rightarrow y'_x = -\frac{f'_x(x, y)}{f'_y(x, y)} = -\frac{\frac{d}{dx}f(x, y)}{\frac{d}{dy}f(x, y)}$$

Based on the information given in the problem, we have:

$$x^3 + y^3 = 16$$

The problem can be solved as follows:

$$y'_x = y' = -\frac{3x^2}{3y^2} = -\frac{x^2}{y^2}$$

$$\Rightarrow y'' = -\frac{2xy^2 - 2yy'x^2}{y^4}$$

$$\xrightarrow{\quad y' = -\dfrac{x^2}{y^2}\quad} y'' = -\frac{2xy^2 - 2y\left(-\frac{x^2}{y^2}\right)x^2}{y^4} = -\frac{2xy^3 + 2x^4}{y^5} = -\frac{2x(y^3 + x^3)}{y^5}$$

$$\xrightarrow{\quad x^3 + y^3 = 16\quad} y'' = -\frac{2x \times 16}{y^5} \Rightarrow y'' = -\frac{32x}{y^5}$$

Choice (4) is the answer.

8.21. From derivative rules, we know that:

$$\begin{cases} y = y(t) \\ x = x(t) \end{cases} \Rightarrow y'_x = \frac{y'_t}{x'_t} = \frac{\frac{d}{dt}y(t)}{\frac{d}{dt}x(t)}$$

Based on the information given in the problem, we have:

$$x = 2 + 3\sin(t)$$

$$y = 3 - 2\cos(t)$$

Hence:

$$y'_x = \frac{y'_t}{x'_t} = \frac{2\sin(t)}{3\cos(t)} = \frac{2}{3}\tan(t)$$

$$t = \frac{\pi}{6} \Rightarrow y'_x = \frac{2}{3} \times \frac{\sqrt{3}}{3} \Rightarrow y'_x = \frac{2\sqrt{3}}{9}$$

Choice (1) is the answer.

8.22. From derivative rules, we know that:

$$\begin{cases} y = y(t) \\ x = x(t) \end{cases} \Rightarrow y'_x = \frac{y'_t}{x'_t} = \frac{\frac{d}{dt}y(t)}{\frac{d}{dt}x(t)}, \quad x'_y = \frac{x'_t}{y'_t} = \frac{\frac{d}{dt}x(t)}{\frac{d}{dt}y(t)}$$

Based on the information given in the problem, we have:

$$x = t^2 + t$$

$$y = t^2 - 2t$$

Therefore:

$$\Rightarrow \begin{cases} y'_x = \dfrac{y'_t}{x'_t} = \dfrac{2t-2}{2t+1} \\ x'_y = \dfrac{x'_t}{y'_t} = \dfrac{2t+1}{2t-2} \end{cases} \xrightarrow{\quad t = -1\quad} \begin{cases} y'_x = \dfrac{-2-2}{-2+1} = 4 \\ x'_y = \dfrac{-2+1}{-2-2} = \dfrac{1}{4} \end{cases} \Rightarrow y'_x + x'_x = 4 + \dfrac{1}{4} \Rightarrow y'_x + x'_x = \dfrac{17}{4}$$

Choice (4) is the answer.

8.23. Based on the information given in the problem, we have:

$$\lim_{h \to 0} \frac{f(x+h) - f(x-h)}{h} = 2\sqrt{x} \qquad (1)$$

From definition of derivative, we know:

$$f'(x) = \lim_{h \to 0} \frac{f(x+h) - f(x)}{h} = \lim_{h \to 0} \frac{f(x) - f(x-h)}{h}$$

The problem can be solved as follows:

$$\lim_{h \to 0} \frac{f(x+h) - f(x-h)}{h} = \cdot \lim_{h \to 0} \frac{f(x+h) - f(x) + f(x) - f(x-h)}{h}$$

$$= \lim_{h \to 0} \frac{f(x+h) - f(x)}{h} + \lim_{h \to 0} \frac{f(x) - f(x-h)}{h} = 2f'(x) \qquad (2)$$

Solving (1) and (2):

$$2f'(x) = 2\sqrt{x} \Rightarrow f'(x) = \sqrt{x} \Rightarrow f'(4) = \sqrt{4} = 2$$

Choice (4) is the answer.

8.24. The function can be simplified as follows:

$$f(x) = x^2 |x| = \begin{cases} x^3 & x \geq 0 \\ -x^3 & x < 0 \end{cases}$$

$$\Rightarrow f(0) = \begin{cases} 0 & x \geq 0 \\ 0 & x < 0 \end{cases} \Rightarrow f(0^-) = f(0^+) \qquad (1)$$

Now, we can determine its first and second derivatives as follows:

$$\Rightarrow f'(x) = \begin{cases} 3x^2 & x \geq 0 \\ -3x^2 & x < 0 \end{cases} \Rightarrow f'(0) = \begin{cases} 0 & x \geq 0 \\ 0 & x < 0 \end{cases} \Rightarrow f'(0^-) = f'(0^+) \qquad (2)$$

$$\Rightarrow f''(x) = \begin{cases} 6x & x \geq 0 \\ -6x & x < 0 \end{cases} \Rightarrow f''(0) = \begin{cases} 0 & x \geq 0 \\ 0 & x < 0 \end{cases} \Rightarrow f''(0^-) = f''(0^+) \qquad (3)$$

From (1) and (2), we can conclude that the first derivative of the function exits. Likewise, from (1), (2), and (3), we can say that the second derivative of the function does exist. Choice (4) is the answer.

8.25. Based on the information given in the problem, we have:

$$f(x) = \begin{cases} \sin^2(x) - \cos(2x) & 0 < x \leq \dfrac{\pi}{4} \\ a\tan(x) + b\sin(2x) & \dfrac{\pi}{4} < x < \dfrac{\pi}{2} \end{cases} \qquad (1)$$

The function is differentiable at $x = \frac{\pi}{4}$. Therefore, we can conclude that:

$$f\left(\left(\frac{\pi}{4}\right)^-\right) = f\left(\left(\frac{\pi}{4}\right)^+\right) \tag{2}$$

$$f'\left(\left(\frac{\pi}{4}\right)^-\right) = f'\left(\left(\frac{\pi}{4}\right)^+\right) \tag{3}$$

Solving (1) and (2):

$$\sin^2\left(\frac{\pi}{4}\right) - \cos\left(2 \times \frac{\pi}{4}\right) = a\tan\left(\frac{\pi}{4}\right) + b\sin\left(2 \times \frac{\pi}{4}\right) \Rightarrow a + b = \frac{1}{2} \tag{4}$$

Solving (1) and (3):

$$2\sin\left(\frac{\pi}{4}\right)\cos\left(\frac{\pi}{4}\right) + 2\sin\left(2 \times \frac{\pi}{4}\right) = a\left(1 + \tan^2\left(\frac{\pi}{4}\right)\right) + 2b\cos\left(2 \times \frac{\pi}{4}\right)$$

$$\Rightarrow 1 + 2 = 2a + 0 \Rightarrow a = \frac{3}{2} \tag{5}$$

Solving (4) and (5):

$$b = -1$$

Choice (1) is the answer.

8.26. Based on the information given in the problem, we have:

$$f(x) = \begin{cases} ax + b & x < -1 \\ x^2 + a & x \geq -1 \end{cases} \tag{1}$$

The function is differentiable everywhere on $\mathbb{R}$ domain including $x = -1$. Hence:

$$f\left((-1)^-\right) = f\left((-1)^+\right) \tag{2}$$

$$f'\left((-1)^-\right) = f'\left((-1)^+\right) \tag{3}$$

Solving (1) and (2):

$$-a + b = 1 + a \Rightarrow -2a + b = 1 \tag{4}$$

Solving (1) and (3):

$$a = -2 \tag{5}$$

Solving (4) and (5):

$$b = -3$$

Choice (4) is the answer.

8.27. From list of derivative rules, we know that:

$$f(x) = \frac{g(x)}{h(x)} \Rightarrow f'(x) = \frac{g'(x)h(x) - h'(x)g(x)}{h^2(x)}$$

Therefore:

$$f(x) = \frac{(2x-1)^2}{2x^2} \Rightarrow f'(x) = \frac{4(2x-1)(2x^2) - 4x(2x-1)^2}{4x^4}$$

$$\Rightarrow f'(x) = \frac{16x^3 - 8x^2 - 16x^3 + 16x^2 - 4x}{4x^4} = \frac{8x^2 - 4x}{4x^4} = \frac{2x-1}{x^3}$$

Choice (2) is the answer.

8.28. From list of derivative rules, we know that:

$$f(x) = u(x)v(x) \Rightarrow f'(x) = u'(x)v(x) + u(x)v'(x)$$

From trigonometry, we have:

$$1 + \tan^2(x) = \frac{1}{\cos^2(x)}$$

Based on the information given in the problem, we have:

$$f(x) = \frac{\sin(x)}{1 + \tan^2(x)} \Rightarrow f(x) = \frac{\sin(x)}{\frac{1}{\cos^2(x)}} = \sin(x)\cos^2(x)$$

$$\Rightarrow f'(x) = \cos(x)\cos^2(x) - \sin(x) \times 2\sin(x)\cos(x) = \cos^3(x) - 2\sin^2(x)\cos(x)$$

$$\Rightarrow f'\left(\frac{\pi}{3}\right) = \left(\frac{1}{2}\right)^3 - 2\left(\frac{\sqrt{3}}{2}\right)^2 \frac{1}{2} = \frac{1}{8} - \frac{3}{4} = -\frac{5}{8}$$

Choice (4) is the answer.

8.29. From list of derivative rules, we know that:

$$f(x) = u(x)v(x) \Rightarrow f'(x) = u'(x)v(x) + u(x)v'(x)$$

$$f(x) = ax^n \Rightarrow f'(x) = anx^{n-1}$$

From trigonometry, we know that:

$$arc\left(\sin\left(\frac{1}{2}\right)\right) = \frac{\pi}{6}$$

Based on the information given in the problem, we have:

$$f(x) = \left(x^2 - 5x + 6\right)arc\left(\sin\left(\frac{1}{x}\right)\right)$$

Therefore:

$$f'(x) = (2x - 5)\text{arc}\left(\sin\left(\frac{1}{x}\right)\right) + (x^2 - 5x + 6)\left(\text{arc}\left(\sin\left(\frac{1}{x}\right)\right)\right)'$$

$$f'(2) = (2 \times 2 - 5)\text{arc}\left(\sin\left(\frac{1}{2}\right)\right) + 0 \times \left(\text{arc}\left(\sin\left(\frac{1}{x}\right)\right)\right)' = (-1) \times \frac{\pi}{6} \Rightarrow f'(2) = -\frac{\pi}{6}$$

As can be seen, we did not need to calculate the value of $\left(\text{arc}\left(\sin\left(\frac{1}{x}\right)\right)\right)'$. Choice (3) is the answer.

8.30. From list of derivative rules, we know that:

$$f(x) = u(x)v(x) \Rightarrow f'(x) = u'(x)v(x) + u(x)v'(x)$$

$$f(x) = ax^n \Rightarrow f'(x) = anx^{n-1}$$

Based on the information given in the problem, we have:

$$f(x) = \left(x^2 + 2x - 3\right) \times \frac{g(x + 2)}{(x^3 + 1)g(2x + 5)} \boxed{?}u(x) \times v(x)$$

The problem can be solved as follows:

$$f'(x) = (2x + 2)\frac{g(x + 2)}{(x^3 + 1)g(2x + 5)} + \left(x^2 + 2x - 3\right)v'(x)$$

$$\Rightarrow f'(x = -3) = (-4) \times \frac{g(-1)}{-26g(-1)} + (0) \times v'(x = -3) = -4 \times \left(-\frac{1}{26}\right)$$

$$\Rightarrow f'(x = -3) = \frac{2}{13}$$

As can be seen, we did not need to calculate the value of $v'(x = -3)$. Choice (4) is the answer.

8.31. Since the line is tangent to the curve, equating their equations and solving them will result in a new equation that will have repeated roots. In other words, the discriminant of the new equation must be zero ($\Delta = 0$).

$$\frac{-1 + x^2}{m + x} = 2x + 1 \Rightarrow -1 + x^2 = 2x^2 + x + 2mx + m \Rightarrow x^2 + (2m + 1)x + m + 1 = 0$$

$$\Delta = 0 \Rightarrow (2m + 1)^2 - 4(m + 1) = 0 \Rightarrow 4m^2 - 3 = 0 \Rightarrow m = \pm\frac{\sqrt{3}}{2}$$

Choice (4) is the answer.

8.32. From list of derivative rules, we know that:

$$f(x) = ax^n \Rightarrow f'(x) = anx^{n-1}$$

Based on the information given in the problem, we have:

$$f(x) = x^4|x|$$

Therefore:

$$f(x) = \begin{cases} x^5, & x \geq 0 \\ -x^5, & x < 0 \end{cases} \Rightarrow f'(x) = \begin{cases} 5x^4, & x > 0 \\ 0, & x = 0 \\ -5x^4, & x < 0 \end{cases} \Rightarrow f''(x) = \begin{cases} 20x^3, & x > 0 \\ 0, & x = 0 \\ -20x^3, & x < 0 \end{cases}$$

$$\Rightarrow f'''(x) = \begin{cases} 60x^2, & x > 0 \\ 0, & x = 0 \\ -60x^2, & x < 0 \end{cases} \Rightarrow f'''(x) = 60x|x|$$

Choice (4) is the answer.

8.33. Based on the information given in the problem, we have:

$$f(x) = \sqrt{x+a} \Rightarrow f'(2) = \frac{1}{4} \tag{1}$$

From list of derivative rules, we know that:

$$f(x) = \sqrt{u(x)} \Rightarrow f'(x) = \frac{u'(x)}{2\sqrt{u(x)}} \tag{2}$$

Solving (1) and (2):

$$\frac{1}{2\sqrt{x+a}}\bigg|_{x=2} = \frac{1}{4} \Rightarrow \frac{1}{2\sqrt{2+a}} = \frac{1}{4} \Rightarrow \sqrt{2+a} = 2 \Rightarrow a = 2$$

Choice (4) is the answer.

8.34. First, we should simplify the function as follows:

$$y(x) = \ln e^{\sqrt{\sin(x)}} = \sqrt{\sin(x)} \tag{1}$$

From list of derivative rules, we know that:

$$f(x) = \sqrt{u(x)} \Rightarrow f'(x) = \frac{u'(x)}{2\sqrt{u(x)}} \tag{2}$$

Solving (1) and (2):

$$\Rightarrow y'(x) = \frac{\cos(x)}{2\sqrt{\sin(x)}} \Rightarrow y'\left(\frac{\pi}{6}\right) = \frac{\cos\left(\frac{\pi}{6}\right)}{2\sqrt{\sin\left(\frac{\pi}{6}\right)}} = \frac{\frac{\sqrt{3}}{2}}{2\sqrt{\frac{1}{2}}} = \frac{\sqrt{6}}{4}$$

Choice (4) is the answer.

8.35. From list of derivative rules, we know that:

$$f(x) = ax^n \Rightarrow f'(x) = anx^{n-1}$$

To determine the maximum value of a function for a given range, we need to calculate the value of the function at its critical points including the extremum points and the beginning and the end of the range.

$$y(x) = x^3 - 3x^2 - 9x + 5 \Rightarrow y'(x) = 3x^2 - 6x - 9$$

To find the extremum points of the function:

$$\Rightarrow y'(x) = 0 \Rightarrow 3x^2 - 6x - 9 = 0 \Rightarrow x^2 - 2x - 3 = 0 \Rightarrow x = 3, -1$$

$x = 3$ is not acceptable because it is out of the range. The value of the function at the critical points can be calculated as follows:

$$y(-2) = (-2)^3 - 3(-2)^2 - 9(-2) + 5 = 3$$

$$y(-1) = (-1)^3 - 3(-1)^2 - 9(-1) + 5 = 10$$

$$y(2) = (2)^3 - 3(2)^2 - 9(2) + 5 = -17$$

Therefore, the maximum value of the function is 10. Choice (2) is the answer.

8.36. From list of derivative rules, we know that:

$$f(x) = \frac{g(x)}{h(x)} \Rightarrow f'(x) = \frac{g'(x)h(x) - h'(x)g(x)}{h^2(x)} \tag{1}$$

First, we need to take its derivative as follows:

$$y(x) = \frac{1 - \sin(x)}{\cos(x)} \Rightarrow y'(x) = \frac{-\cos(x)\cos(x) - (-\sin(x))(1 - \sin(x))}{\cos^2(x)}$$

$$y'(x) = \frac{-\cos^2(x) + \sin(x) - \sin^2(x)}{\cos^2(x)} = \frac{-1 + \sin(x)}{\cos^2(x)}$$

The $y'(x)$ is always nonpositive because $\sin(x) \le 1$. Hence, the function is a descending function. Choice (2) is the answer.

8.37. Based on the information given in the problem, we have:

$$f(x) = \left(\sqrt{1 + x^2} - x\right)^5, \quad g(x) = \frac{1}{\left(\sqrt{1 + x^2} + x\right)^5} \tag{1}$$

From list of derivative rules, we know that:

$$\left(\frac{f(x)}{g(x)}\right)' = \frac{f'(x)g(x) - g'(x)f(x)}{(g(x))^2} \Rightarrow f'(x)g(x) - g'(x)f(x) = \left(\frac{f(x)}{g(x)}\right)'(g(x))^2 \tag{2}$$

Solving (1) and (2):

$$f'(x)g(x) - g'(x)f(x) = \left(\frac{\left(\sqrt{1+x^2} - x\right)^5}{\frac{1}{\left(\sqrt{1+x^2}+x\right)^5}}\right)' \left(\frac{1}{\left(\sqrt{1+x^2}+x\right)^5}\right)^2$$

$$= \left(\left(1 + x^2 - x^2\right)^5\right)' \frac{1}{\left(\sqrt{1+x^2}+x\right)^{10}} = (1)' \times \frac{1}{\left(\sqrt{1+x^2}+x\right)^{10}} = 0 \times \frac{1}{\left(\sqrt{1+x^2}+x\right)^{10}} = 0$$

Choice (2) is the answer.

8.38. From list of derivative rules, we know that:

$$\frac{d}{dx}\left(\ln f(x)\right) = \frac{f'(x)}{f(x)} \tag{1}$$

$$\frac{d}{dx}\left(\sqrt{f(x)}\right) = \frac{f'(x)}{2\sqrt{f(x)}} \tag{1}$$

In addition, we know that:

$$\ln 1 = 0$$

$$\ln \sqrt[m]{f(x)} = \frac{1}{m}\ln f(x)$$

$$\ln(f(x)g(x)) = \ln(f(x)) + \ln(g(x))$$

The problem can be solved as follows:

$$y(x) = \frac{1}{\sqrt{1+x^2}\left(x + \sqrt{1+x^2}\right)}$$

$$\xrightarrow{\ln} \ln y(x) = \ln 1 - \ln\left(\sqrt{1+x^2}\left(x + \sqrt{1+x^2}\right)\right) = 0 - \ln\sqrt{1+x^2} - \ln\left(x + \sqrt{1+x^2}\right)$$

$$\Rightarrow \ln y(x) = -\frac{1}{2}\ln\left(1+x^2\right) - \ln\left(x + \sqrt{1+x^2}\right)$$

$$\xrightarrow{\frac{d}{dx}} \frac{y'(x)}{y(x)} = -\frac{1}{2}\frac{2x}{1+x^2} - \frac{1 + \frac{2x}{2\sqrt{1+x^2}}}{x + \sqrt{1+x^2}}$$

$$\Rightarrow y'(x) = y(x)\left(-\frac{1}{2}\frac{2x}{1+x^2} - \frac{1 + \frac{2x}{2\sqrt{1+x^2}}}{x + \sqrt{1+x^2}}\right)$$

$$\Rightarrow y'(x) = \left(\frac{1}{\sqrt{1+x^2}\left(x + \sqrt{1+x^2}\right)}\right)\left(-\frac{1}{2}\frac{2x}{1+x^2} - \frac{1 + \frac{2x}{2\sqrt{1+x^2}}}{x + \sqrt{1+x^2}}\right)$$

Now, for $x = \frac{4}{3}$, we have:

$$y'\left(\frac{4}{3}\right) = -\frac{27}{125}$$

Choice (1) is the answer.

8.39. Since the tangent lines are parallel to x-axis, their slope angles must be zero. Therefore:

$$y = x^3 - 6x + 12 \Rightarrow y' = 3x^2 - 6 = 0 \Rightarrow x = \sqrt{2}, -\sqrt{2}$$

$$x_1 = \sqrt{2} \Rightarrow y_1 = \left(\sqrt{2}\right)^3 - 6\sqrt{2} + 12 = 2\sqrt{2} - 6\sqrt{2} + 12 = -4\sqrt{2} + 12$$

$$x_2 = -\sqrt{2} \Rightarrow y_2 = \left(-\sqrt{2}\right)^3 - 6\left(-\sqrt{2}\right) + 12 = -2\sqrt{2} + 6\sqrt{2} + 12 = 4\sqrt{2} + 12$$

$$y_2 - y_1 = \left(4\sqrt{2} + 12\right) - \left(-4\sqrt{2} + 12\right) \Rightarrow y_2 - y_1 = 8\sqrt{2}$$

Choice (4) is the answer.

8.40. Based on the information given in the problem, we have:

$$f(x) = \begin{cases} \dfrac{(x+1)^5}{|x+1|} & x \neq -1 \\ 0 & x = -1 \end{cases}$$

This problem can be solved by using the definition of derivative of a function as follows:

$$f'(x_0) = \lim_{x \to x_0} \frac{f(x) - f(x_0)}{x - x_0}$$

$$\Rightarrow f'(-1) = \lim_{x \to (-1)} \frac{\frac{(x+1)^5}{|x+1|} - 0}{x - (-1)} = \lim_{x \to (-1)} \frac{\frac{(x+1)^5}{|x+1|}}{x+1} = \lim_{x \to (-1)} \frac{(x+1)^4}{|x+1|} = \lim_{x \to (-1)} |x+1|^3 = 0$$

Choice (1) is the answer.

8.41. From list of derivative rules, we know that:

$$f(x) = \sqrt{u(x)} \Rightarrow f'(x) = \frac{u'(x)}{2\sqrt{u(x)}} \qquad (1)$$

The distance of the point from the origin can be calculated as follows:

$$D(x) = \sqrt{(x-0)^2 + \left(\sqrt{x+8} - 0\right)^2} = \sqrt{x^2 + x + 8} \qquad$$

In addition, the changing rate of the distance can be determined as follows:

$$D'(x) = \frac{d}{dx} D(x) = \frac{d}{dx} \sqrt{x^2 + x + 8} \qquad (2)$$

Solving (1) and (2):

$$D'(x) = \frac{2x+1}{2\sqrt{x^2+x+8}} \Rightarrow D'(7) = \frac{2 \times 7 + 1}{2\sqrt{7^2+7+8}} = \frac{15}{16}$$

Choice (1) is the answer.

8.42. From list of derivative rules, we know that:

$$f(x) = g(h(x)) \Rightarrow f'(x) = h'(x)g'(h(x))$$

$$f(x) = \sqrt{u(x)} \Rightarrow f'(x) = \frac{u'(x)}{2\sqrt{u(x)}}$$

The problem should be solved using the definition of derivative of a function as follows:

$$f'(x_0) = \lim_{x \to x_0} \frac{f(x) - f(x_0)}{x - x_0} \tag{1}$$

Based on the information given in the problem, we have:

$$\lim_{x \to 2} \frac{f(x) - f(2)}{x - 2} = -\frac{1}{3} \tag{2}$$

Solving (1) and (3):

$$f'(2) = -\frac{1}{3} \tag{3}$$

Therefore:

$$\frac{d}{dx}\left(f\left(\sqrt{|-x|+3}\right)\right)\Big|_{x=-1} = \frac{d}{dx}\left(f(\sqrt{-x+3})\right)\Big|_{x=-1} \Rightarrow \frac{-1}{2\sqrt{-x+3}}f'\left(\sqrt{-x+3}\right)\Big|_{x=-1} = -\frac{1}{4}f'(2) \tag{4}$$

Solving (3) and (4):

$$\frac{d}{dx}\left(f\left(\sqrt{|-x|+3}\right)\right)\Big|_{x=-1} = -\frac{1}{4} \times \left(-\frac{1}{3}\right) = \frac{1}{12}$$

Choice (2) is the answer.

8.43. Based on the information given in the problem, we have:

$$f(x) = \frac{(x+1)h(x)}{(2x+1)h(2x+1)}, \quad h(-1) \neq 0 \tag{1}$$

The derivative of this function should be solved by using the definition of derivative of a function as follows:

$$f'(x_0) = \lim_{x \to x_0} \frac{f(x) - f(x_0)}{x - x_0} \Rightarrow f'(-1) = \lim_{x \to -1} \frac{f(x) - f(-1)}{x - (-1)} \tag{2}$$

Solving (1) and (2):

$$f'(-1) = \lim_{x \to -1} \frac{\frac{(x+1)h(x)}{(2x+1)h(2x+1)} - \frac{(-1+1)h(-1)}{(-2+1)h(-2+1)}}{x - (-1)} = \lim_{x \to -1} \frac{\frac{(x+1)h(x)}{(2x+1)h(2x+1)} - 0}{x+1}$$

$$f'(-1) = \lim_{x \to -1} \frac{h(x)}{(2x+1)h(2x+1)} = \frac{h(-1)}{(-1)h(-1)} = -1$$

Choice (2) is the answer.

8.44. Based on the information given in the problem, the width of the extremum point is as follows:

$$y(x_M) = \frac{3}{4} \tag{1}$$

To determine the extremum points of a function, we need to find the roots of the derivative of the function as follows:

$$f'(x) = 0 \tag{2}$$

$$f(x) = \cos^2(x) + \sqrt{3}\sin(x) + a \tag{3}$$

$$\Rightarrow f'(x) = -2\sin(x)\cos(x) + \sqrt{3}\cos(x) = \cos(x)\left(-2\sin(x) + \sqrt{3}\right) \tag{4}$$

Solving (2) and (4):

$$\cos(x)\left(-2\sin(x) + \sqrt{3}\right) = 0 \Rightarrow \begin{cases} \cos(x) = 0 & (5) \\ \sin(x) = \dfrac{\sqrt{3}}{2} & (6) \end{cases}$$

There is no answer for equation (5) in the range of $0 < x < \frac{\pi}{2}$. However, $x = \frac{\pi}{3}$ is only answer for equation (6).

Therefore, using $x_M = \frac{\pi}{3}$ and (1) in (3), we have:

$$\frac{3}{4} = \cos^2\left(\frac{\pi}{3}\right) + \sqrt{3}\sin\left(\frac{\pi}{3}\right) + a \Rightarrow \frac{3}{4} = \frac{1}{4} + \frac{3}{2} + a \Rightarrow a = -1$$

Choice (4) is the answer.

8.45. From list of derivative rules, we know that:

$$\frac{d}{dx}(\ln f(x)) = \frac{f'(x)}{f(x)} \tag{1}$$

$$\frac{d}{dx}(u(x)v(x)) = u'(x)v(x) + u(x)v'(x) \tag{2}$$

Moreover, we know that:

$$\ln a^b = a \ln b$$

$$\ln e = 1$$

The problem can be solved as follows:

$$y(x) = x^x$$

$$\xrightarrow{\ \ln\ } \ln y(x) = \ln x^x \Rightarrow \ln y(x) = x \ln x$$

$$\xrightarrow{\ \frac{d}{dx}\ } \frac{y'(x)}{y(x)} = \ln x + x\left(\frac{1}{x}\right) = \ln x + 1$$

$$\Rightarrow y'(x) = y(x)(\ln x + 1) \Rightarrow y'(x) = x^x(\ln x + 1)$$

$$\Rightarrow y'(e) = e^e(\ln e + 1) = 2e^e$$

Choice (3) is the answer.

8.46. From list of derivative rules, we know that:

$$\frac{d}{dx}(\ln f(x)) = \frac{f'(x)}{f(x)} \tag{1}$$

$$\frac{d}{dx}((f(x))^n) = nf'(x)(f(x))^{n-1} \tag{2}$$

Also, we know that:

$$\ln f(x)^{g(x)} = g(x)\ln f(x)$$

The problem can be solved as follows:

$$y(x) = x^{\ln x}$$

$$\xrightarrow{\ \ln\ } \ln y(x) = \ln x^{\ln x} = \ln x \times \ln x \Rightarrow \ln y(x) = (\ln x)^2$$

$$\xrightarrow{\ \frac{d}{dx}\ } \frac{y'(x)}{y(x)} = 2\left(\frac{1}{x}\right)\ln x$$

$$\Rightarrow y'(x) = \frac{2x^{\ln x}\ln x}{x}$$

Choice (3) is the answer.

8.47. From list of derivative rules, we know that:

$$\frac{d}{dx}\left(\frac{f(x)}{g(x)}\right) = \frac{f'(x)g(x) - f(x)g'(x)}{(g(x))^2} \tag{1}$$

We can find a formula for the *n-th* derivative of the function below as follows:

$$y(x) = \frac{1}{x}$$

$$\xrightarrow{\frac{d}{dx}} y'(x) = \frac{0 - 1 \times 1}{x^2} = -\frac{1}{x^2} = (-1)^1 \frac{1}{x^{1+1}}$$

$$\xrightarrow{\frac{d^2}{dx^2}} y''(x) = -\frac{5 - 2x \times 1}{(x^2)^2} = \frac{2}{x^3} = (-1)^2 \frac{2 \times 1}{x^{2+1}}$$

$$\xrightarrow{\frac{d^3}{dx^3}} y'''(x) = \frac{5 - 3x^2 \times 2}{(x^3)^3} = \frac{-6}{x^4} = (-1)^3 \frac{3 \times 2 \times 1}{x^{3+1}}$$

$$\vdots$$

$$\xrightarrow{\frac{d^n}{dx^n}} y^{(n)}(x) = (-1)^n \frac{n!}{x^{n+1}}$$

Choice (4) is the answer.

8.48. The equation of a line which is tangent on a curve at the point of (x_0, y_0), located on the curve, can be calculated as follows:

$$y - y_0 = m(x - x_0)$$

Moreover, the equation of a line which is perpendicular on a curve at the point of (x_0, y_0), located on the curve, can be calculated as follows:

$$y - y_0 = m'(x - x_0)$$

where m and m' are the slope of the tangent and perpendicular lines. In addition, we have:

$$m = y'(x_0)$$

$$m' = -\frac{1}{m}$$

Therefore, first, we need to determine the first derivative of $y(x)$.

$$y(x) = x^{2x}$$

$$\xrightarrow{\ln} \ln y(x) = \ln x^{2x} = 2x \ln x$$

$$\xrightarrow{\frac{d}{dx}} \frac{y'(x)}{y(x)} = 2\ln x + 2$$

$$\Rightarrow y'(x) = y(x)(2\ln x + 2) = x^{2x}(2\ln x + 2)$$

The slope of the tangent line:

$$m = y'(x_0 = 1) = 1^2(2\ln 1 + 2) = 2$$

The slope of the perpendicular line:

$$m' = -\frac{1}{m} = -\frac{1}{2}$$

The equation of the perpendicular line:

$$y - 1 = -\frac{1}{2}(x - 1)$$

$$\Rightarrow y = -\frac{1}{2}x + \frac{3}{2} \Rightarrow x + 2y - 3 = 0$$

Choice (1) is the answer.

In this problem, the rules below were used.

$$\ln a^b = b \ln a$$

$$\frac{d}{dx}(\ln f(x)) = \frac{f'(x)}{f(x)}$$

8.49. The angle between the right and left tangent lines of a function can be determined as follows:

$$\theta = \pi - \tan^{-1}\left|\frac{m - m'}{1 + mm'}\right|$$

where m and m' are the slope of the right and left tangent lines.

For the following function at the given point of $x_0 = 1$, we have:

$$f(x) = \begin{cases} x^3 & x > 1 \\ \sqrt{x} & x \le 1 \end{cases}$$

$$m = f'(1^+) = 3x^2\big|_{x_0 = 1^+} = 3$$

$$m' = f'(1^-) = \frac{1}{2\sqrt{x}}\bigg|_{x_0 = 1^-} = \frac{1}{2}$$

Therefore:

$$\theta = \pi - \tan^{-1}\left|\frac{3 - \frac{1}{2}}{1 + 3 \times \frac{1}{2}}\right| = \pi - \tan^{-1}(1) = \pi - \frac{\pi}{4}$$

$$\theta = \frac{3\pi}{4}$$

Choice (4) is the answer.

In this problem, the rules below were used.

$$m = f'(x_0)$$

$$\tan^{-1}(1) = \frac{\pi}{4}$$

References

1. Rahmani-Andebili, M. (2021). Calculus – Practice Problems, Methods, and Solutions, Springer Nature, 2021.
2. Rahmani-Andebili, M. (2021). Precalculus – Practice Problems, Methods, and Solutions, Springer Nature, 2021.

Abstract

In this chapter, the basic and advanced problems of definite and indefinite integrals are presented. The subjects include definite integrals, indefinite integrals, substitution rule for integrals, integration techniques, integration by parts, integrals involving trigonometric functions, trigonometric substitutions, integration using partial fractions, integrals involving roots, and integrals involving quadratics. To help students study the chapter in the most efficient way, the problems are categorized in different levels based on their difficulty levels (easy, normal, and hard) and calculation amounts (small, normal, and large). Moreover, the problems are ordered from the easiest problem with the smallest computations to the most difficult problems with the largest calculations.

9.1. Calculate the value of the indefinite integral below [1, 2].

$$I = \int (3x + 5)^{17} dx$$

Difficulty level   ● Easy   ○ Normal   ○ Hard
Calculation amount   ● Small   ○ Normal   ○ Large

1) $(3x + 5)^{18} + c$

2) $\dfrac{(3x + 5)^{18}}{54} + c$

3) $\dfrac{(3x + 5)^{17}}{3} + c$

4) $\dfrac{(3x + 5)^{18}}{18} + c$

9.2. Calculate the value of the following indefinite integral.

$$I = \int \cos(1 + \pi x) dx$$

Difficulty level   ● Easy   ○ Normal   ○ Hard
Calculation amount   ● Small   ○ Normal   ○ Large

1) $\sin(1 + \pi x) + c$

2) $\dfrac{1}{1 + \pi} \sin(1 + \pi x) + c$

3) $\cos(1 + \pi x) + c$

4) $\dfrac{1}{\pi} \sin(1 + \pi x) + c$

9.3. If $F(x) = \int f(x)dx$, calculate the value of $\int f(ax + b)dx$.

1) $aF(ax + b)$

2) $\dfrac{1}{a}F(x)$

3) $aF(x)$

4) $\dfrac{1}{a}F(ax + b)$

9.4. Calculate the value of the definite integral below.

$$\int_1^2 \frac{x + 4}{x^3}\,dx$$

1) 1
2) 2
3) 3
4) 4

9.5. Solve the following indefinite integral:

$$\int \left(e^x + 2xe^{x^2} \right) dx$$

1) $-e^x - e^{x^2} + c$

2) $-e^x + e^{x^2} + c$

3) $e^x - e^{x^2} + c$

4) $e^x + e^{x^2} + c$

9.6. Calculate the value of $f''(1)$ if we know that:

$$f(x) = \int \left(x^3 + 5x \right) dx$$

1) 2
2) 4
3) 6
4) 8

9.7. Calculate the value of $f''\left(\dfrac{\pi}{2}\right)$ if $f(x) = \int \cos^3(x)dx$.

1) 0
2) 1

3) -1

4) $\dfrac{3\sqrt{2}}{2}$

9.8. Calculate the value of the definite integral below.

$$\int_{-2}^{-1} \frac{x^3 + x^2 - 1}{x^2} dx$$

Difficulty level   ● Easy   ○ Normal   ○ Hard
Calculation amount   ○ Small   ● Normal   ○ Large

1) 1

2) -1

3) $\dfrac{1}{2}$

4) $-\dfrac{1}{2}$

9.9. Solve the indefinite integral below.

$$\int \frac{x - 2}{\sqrt{x}} dx$$

Difficulty level   ● Easy   ○ Normal   ○ Hard
Calculation amount   ○ Small   ● Normal   ○ Large

1) $\dfrac{2}{3}\sqrt{x}(x + 6) + c$

2) $\dfrac{2}{3}\sqrt{x}(x - 6) + c$

3) $\dfrac{1}{3}\sqrt{x}(x + 6) + c$

4) $\dfrac{2}{3}\sqrt{x}(x - 6) + c$

9.10. Calculate the value of the following definite integral:

$$I = \int_0^1 \frac{x^2}{(x^3 + 1)^4} dx$$

Difficulty level   ● Easy   ○ Normal   ○ Hard
Calculation amount   ○ Small   ● Normal   ○ Large

1) $\dfrac{5}{36}$

2) $\dfrac{7}{36}$

3) $\dfrac{5}{72}$

4) $\dfrac{7}{72}$

9.11. Calculate the value of the following indefinite integral.

$$I = \int \frac{dx}{\cos^2 x \sqrt{1 + \tan x}}$$

Difficulty level   ○ Easy   ● Normal   ○ Hard
Calculation amount   ● Small   ○ Normal   ○ Large

1) $\dfrac{1}{\cos x} + \tan x + c$

2) $\dfrac{1}{\cos x} + \cot gx + c$

3) $2\sqrt{1 + \tan x} + c$

4) $2(1 + \tan x) + c$

9.12. Calculate the value of the indefinite integral below.

$$I = \int \frac{dx}{x\sqrt{\ln x}}$$

Difficulty level   ○ Easy   ● Normal   ○ Hard
Calculation amount   ● Small   ○ Normal   ○ Large

1) $\dfrac{2}{\sqrt{\ln x}} + c$

2) $\dfrac{\sqrt{\ln x}}{x} + c$

3) $2\sqrt{\ln x} + c$

4) $\ln(\ln \sqrt{x})$

9.13. Calculate the integral of the function below for the range of $\infty < x < +\infty$.

$$f(x) = \frac{1}{x^2 + 4}$$

Difficulty level   ○ Easy   ● Normal   ○ Hard
Calculation amount   ● Small   ○ Normal   ○ Large

1) $\dfrac{\pi}{2}$

2) π

3) $\dfrac{3\pi}{2}$

4) 2π

9.14. Calculate the value of the definite integral below.

$$\int_1^4 \frac{x - \sqrt{x}}{\sqrt{x}} dx$$

Difficulty level   ○ Easy   ● Normal   ○ Hard
Calculation amount   ● Small   ○ Normal   ○ Large

1) $\dfrac{1}{3}$

2) $\dfrac{2}{3}$

3) $\dfrac{4}{3}$

4) $\dfrac{5}{3}$

9.15. If the primary function of $f(x)$ is equal to $\dfrac{x^3}{6}$, determine the first derivate of $f\left(\dfrac{1}{x}\right)$ with respect to x.

Difficulty level   ○ Easy   ● Normal   ○ Hard
Calculation amount   ● Small   ○ Normal   ○ Large

1) $-\dfrac{1}{x^3}$

2) $-\dfrac{x}{6}$

3) $-\dfrac{1}{2}$

4) $\dfrac{1}{x^3}$

9.16. Calculate the value of $F^{'}(\lambda = 0)$ if:

$$F(\lambda) = \int_0^\lambda \frac{1}{x^4 + 2} dx$$

Difficulty level   ○ Easy   ● Normal   ○ Hard
Calculation amount   ● Small   ○ Normal   ○ Large

1) 1

2) $\dfrac{1}{2}$

3) $\dfrac{1}{3}$

4) 0

9.17. Calculate the value of the definite integral below.

$$\int_{-1}^1 \frac{x^2}{1 + x^2} \arc(\tan(x)) dx$$

Difficulty level   ○ Easy   ● Normal   ○ Hard
Calculation amount   ● Small   ○ Normal   ○ Large   .

1) 0

2) 1

3) $\dfrac{\pi}{4}$

4) $\dfrac{\pi}{2}$

9.18. Calculate the integral of the function below for the range of $-\dfrac{1}{2} < x < \dfrac{1}{2}$.

$$f(x) = \frac{1}{\sqrt{1 - x^2}}$$

Difficulty level   ○ Easy   ● Normal   ○ Hard
Calculation amount   ● Small   ○ Normal   ○ Large

1) $\dfrac{\pi}{6}$

2) $\dfrac{\pi}{3}$

3) $-\dfrac{\pi}{6}$

4) $-\dfrac{\pi}{3}$

9.19. Calculate the value of the definite integral below.

$$\int_0^1 \frac{1}{\sqrt{2x - x^2}}\,dx$$

Difficulty level   ○ Easy   ● Normal   ○ Hard
Calculation amount   ● Small   ○ Normal   ○ Large

1) 0

2) $\dfrac{\pi}{4}$

3) $\dfrac{\pi}{2}$

4) π

9.20. Calculate the value of the definite integral below.

$$\int_{-1}^1 \left(x^2 + 1\right)\left(x^3 + 3x\right)dx$$

Difficulty level   ○ Easy   ● Normal   ○ Hard
Calculation amount   ● Small   ○ Normal   ○ Large

1) 21

2) 0

3) −11

4) 2

9.21. Solve the following indefinite integral:

$$\int \frac{\sin(x)}{1 + \cos(\cos(x))}\,dx$$

Difficulty level   ○ Easy   ● Normal   ○ Hard
Calculation amount   ○ Small   ● Normal   ○ Large

1) $-\tan\left(\dfrac{1}{2}\cos(x)\right) + c$

2) $\tan(\cos(x)) + c$

3) $-\tan\left(\cos(x)\right) + c$

4) $\tan\left(\dfrac{1}{2}\cos(x)\right) + c$

9.22. Determine the function of a curve that passes from the point of (3, 4) and its derivative is $-\dfrac{x}{y}$.

Difficulty level   ○ Easy   ● Normal   ○ Hard
Calculation amount   ○ Small   ● Normal   ○ Large

1) $2x^2 + y^2 = 34$

2) $x^2 + y^2 = 16$

3) $y^2 = 4x + 4$

4) $x^2 + y^2 = 25$

9.23. Solve the following indefinite integral:

$$\int \frac{x}{\sqrt{x-1}} dx$$

Difficulty level   ○ Easy   ● Normal   ○ Hard

Calculation amount   ○ Small   ● Normal   ○ Large

1) $\frac{2}{3}(x-1)^{\frac{3}{2}} - 2(x-1)^{\frac{1}{2}} + c$

2) $\frac{2}{3}(x-1)^{\frac{3}{2}} + 2(x-1)^{\frac{1}{2}} + c$

3) $\frac{1}{3}(x-1)^{\frac{3}{2}} - 2(x-1)^{\frac{1}{2}} + c$

4) $-\frac{1}{3}(x-1)^{\frac{3}{2}} - 2(x-1)^{\frac{1}{2}} + c$

9.24. Calculate the value of the definite integral below.

$$\int_{-2}^{5} |x-3| dx$$

Difficulty level   ○ Easy   ● Normal   ○ Hard

Calculation amount   ○ Small   ● Normal   ○ Large

1) $\frac{25}{2}$

2) $\frac{27}{2}$

3) $\frac{29}{2}$

4) $\frac{31}{2}$

9.25. Calculate the value of the definite integral below.

$$\int_{1}^{2} \frac{x(x+1)^2 + 2}{(x+1)^2} dx$$

Difficulty level   ○ Easy   ● Normal   ○ Hard

Calculation amount   ○ Small   ● Normal   ○ Large

1) $\frac{12}{5}$

2) $\frac{9}{5}$

3) $\frac{11}{6}$

4) $\frac{7}{4}$

9.26. Solve the following indefinite integral:

$$\int \frac{1}{1 + e^x} \, dx$$

Difficulty level   ○ Easy   ● Normal   ○ Hard
Calculation amount   ○ Small   ● Normal   ○ Large
1) $x + \ln(1 + e^x) + c$
2) $x - \ln(1 + e^x) + c$
3) $\frac{1}{2}x^2 + \ln(1 + e^x) + c$
4) $\frac{1}{2}x^2 - \ln(1 + e^x) + c$

9.27. Determine the function of a curve that passes from the point of (1, 1) and its derivative is as follows:

$$y' = \frac{x + 1}{1 - y}$$

Difficulty level   ○ Easy   ● Normal   ○ Hard
Calculation amount   ○ Small   ● Normal   ○ Large
1) $x^2 + y^2 + 2x - 2y - 2 = 0$
2) $x^2 - y^2 + 4x - 4y + 1 = 0$
3) $x^2 + y^2 - 2x + 2y - 2 = 0$
4) $x^2 - y^2 + 3x - 2y - 1 = 0$

9.28. What is the function of a curve that passes from the point of (1, 1) and the relation below holds?

$$y' = \frac{3x}{2y}$$

Difficulty level   ○ Easy   ● Normal   ○ Hard
Calculation amount   ○ Small   ● Normal   ○ Large
1) $2y^2 - 3x^2 + 1 = 0$
2) $y^2 - 2x^2 + 1 = 0$
3) $2y^2 + x^2 - 3 = 0$
4) $2y^2 - x^2 - 1 = 0$

9.29. In the equation below, determine the value of A.

$$\int \frac{3x}{\sqrt{x^2 + 1}} \, dx = A\sqrt{x^2 + 1} + c$$

Difficulty level   ○ Easy   ● Normal   ○ Hard
Calculation amount   ○ Small   ● Normal   ○ Large
1) $\frac{1}{2}$
2) 1
3) $\frac{3}{2}$
4) 3

9.30. Calculate the value of the definite integral below.

$$\int_{-1}^{2} |x|\,dx$$

Difficulty level   ○ Easy   ● Normal   ○ Hard
Calculation amount   ○ Small   ● Normal   ○ Large

1) $\dfrac{3}{2}$

2) $\dfrac{5}{2}$

3) $\dfrac{7}{2}$

4) $\dfrac{9}{2}$

9.31. Calculate the value of the following definite integral.

$$\int_{0}^{\frac{\pi}{2}} \sin^{2}(x)\,dx$$

Difficulty level   ○ Easy   ● Normal   ○ Hard
Calculation amount   ○ Small   ● Normal   ○ Large

1) 0

2) $\dfrac{\pi}{2}$

3) $\dfrac{\pi}{4}$

4) $\dfrac{\pi}{8}$

9.32. Calculate the value of the definite integral below.

$$\int_{0}^{\frac{3\pi}{4}} \left(\tan^{5}(x) + \tan^{7}(x)\right)dx$$

Difficulty level   ○ Easy   ● Normal   ○ Hard
Calculation amount   ○ Small   ● Normal   ○ Large

1) 0

2) $\dfrac{1}{2}$

3) $\dfrac{1}{3}$

4) $\dfrac{1}{6}$

9.33. Which one of the points below is on a curve that passes from the point of $(\pi, 1)$ and $y' = y^2 \cos(x)$ holds for that?
Difficulty level   ○ Easy   ● Normal   ○ Hard
Calculation amount   ○ Small   ● Normal   ○ Large

1) $\left(\dfrac{3\pi}{2}, 2\right)$

2) $\left(\dfrac{\pi}{2}, -1\right)$

3) $\left(\dfrac{\pi}{2}, 1\right)$

4) $(0, 1)$

9.34. Calculate the value of the definite integral of I_1 if $I_2 = m$.

$$I_1 = \int_3^5 \dfrac{3x}{x-2}\,dx$$

$$I_2 = \int_3^5 \dfrac{1}{x-2}\,dx$$

Difficulty level   ○ Easy   ● Normal   ○ Hard
Calculation amount   ○ Small   ● Normal   ○ Large

1) $m + 2$

2) $4m - 6$

3) $6m + 6$

4) $6m - 4$

9.35. Calculate the value of the definite integral below.

$$\int_2^3 \dfrac{x}{x^2 - 1}\,dx$$

Difficulty level   ○ Easy   ● Normal   ○ Hard
Calculation amount   ○ Small   ● Normal   ○ Large

1) $\ln\left(\dfrac{8}{3}\right)$

2) $\text{arc}\left(\sin\left(\dfrac{2\sqrt{3}}{5}\right)\right)$

3) $\text{arc}\left(\tan\left(\dfrac{3}{2}\right)\right)$

4) $\ln\left(\sqrt{\dfrac{8}{3}}\right)$

9.36. Calculate the value of the definite integral of $\int_0^4 f'(x)\,dx$ if we have $f(x) = \int_a^x \sqrt{t}\,dt$.

Difficulty level   ○ Easy   ● Normal   ○ Hard
Calculation amount   ○ Small   ● Normal   ○ Large

1) $\dfrac{8}{3}$

2) $\dfrac{16}{3}$

3) $-\dfrac{8}{3}$

4) $-\dfrac{16}{3}$

9.37. Solve the following indefinite integral:

$$\int 8\left(\tan^6(x) + \tan^8(x)\right)dx$$

1) $\tan^7(x) + c$

2) $\dfrac{1}{7}\tan^8(x) + c$

3) $\dfrac{8}{5}\tan^5(x) + c$

4) $\dfrac{8}{7}\tan^7(x) + c$

9.38. Calculate the value of the definite integral below.

$$\int_{-1}^{1} (x+1)\left(x^2 + 2x + 3\right)dx$$

1) 6
2) 4
3) 8
4) 10

9.39. Calculate the value of the definite integral below.

$$\int_{\frac{1}{2}}^{1} \left[\frac{1}{x}\right]\frac{1}{x^3}\,dx$$

1) 1
2) 2
3) $\dfrac{1}{2}$
4) $\dfrac{3}{2}$

9.40. Calculate the value of y'_x if we have:

$$y = u + v, \quad u = \int_{1}^{x^2} \frac{\sin(t)}{t}\,dt, \quad v = \int_{x^2}^{1} \frac{\sin(u)}{u}\,du$$

1) $\dfrac{4\sin(x^2)}{x}$

2) $-\dfrac{4\sin(x^2)}{x}$

3) 0
4) 1

9.41. Calculate the value of $f(3x + 2)$ if we have:

$$h(x) = \int f'(3x + 2)dx, \quad h(0) = 1, \quad f(2) = 3$$

Difficulty level   ○ Easy   ● Normal   ○ Hard

Calculation amount   ○ Small   ● Normal   ○ Large

1) $h(x) - 1$

2) $2h(x) + 1$

3) $3h(x)$

4) $3h(x) - 1$

9.42. A curve is tangent to $y = x$ in the origin and its second derivative is $2x + 1$. Which one of the points below is on the curve?

Difficulty level   ○ Easy   ● Normal   ○ Hard

Calculation amount   ○ Small   ○ Normal   ● Large

1) $\left(1, \dfrac{11}{6}\right)$

2) $\left(1, \dfrac{13}{6}\right)$

3) $\left(2, \dfrac{11}{6}\right)$

4) $\left(2, \dfrac{13}{6}\right)$

9.43. Solve the indefinite integral of $\int \sin(2x)\cos(4x)dx$.

Difficulty level   ○ Easy   ● Normal   ○ Hard

Calculation amount   ○ Small   ○ Normal   ● Large

1) $\dfrac{1}{3}\cos^3(2x) - \dfrac{1}{2}\cos(2x) + c$

2) $-\dfrac{1}{3}\cos^3(2x) + \dfrac{1}{2}\cos(2x) + c$

3) $\dfrac{1}{3}\cos^3(2x) + \dfrac{1}{2}\cos(2x) + c$

4) $-\dfrac{1}{3}\cos^3(2x) - \dfrac{1}{2}\cos(2x) + c$

9.44. Calculate the value of the following indefinite integral.

$$I = \int \frac{e^{\arc \tan x}}{1 + x^2}dx$$

Difficulty level   ○ Easy   ○ Normal   ● Hard

Calculation amount   ● Small   ○ Normal   ○ Large

1) $e^{\arc \tan x} + c$

2) $2e^{\arc \tan x} + c$

3) $\dfrac{1}{2}e^{\arc \tan x} + c$

4) $\arc \tan(e^x + 1) + c$

9.45. Calculate the value of the definite integral below.

$$\int_{-1}^{1} \left[\frac{x}{3}\right]dx$$

Difficulty level   ○ Easy   ○ Normal   ● Hard
Calculation amount   ● Small   ○ Normal   ○ Large
1) 0
2) 1
3) −1
4) −3

9.46. What is the function of a curve that passes from the point of (1, 2) and the relation of $xy' + y = 1$ holds.

Difficulty level   ○ Easy   ○ Normal   ● Hard
Calculation amount   ● Small   ○ Normal   ○ Large

1) $y = 1 + \dfrac{1}{x}$

2) $y = 2 - \dfrac{1}{x}$

3) $y = \dfrac{3}{x} - 1$

4) $y = \dfrac{3}{x} + 1$

9.47. Solve the indefinite integral below.

$$\int \frac{f'(\sqrt[3]{x})}{\sqrt[3]{x^2}}\,dx$$

Difficulty level   ○ Easy   ○ Normal   ● Hard
Calculation amount   ● Small   ○ Normal   ○ Large

1) $\dfrac{1}{3}f(\sqrt[3]{x}) + c$

2) $\dfrac{2}{3}f(\sqrt[3]{x}) + c$

3) $f(\sqrt[3]{x}) + c$

4) $3f(\sqrt[3]{x}) + c$

9.48. Determine the value of the following definite integral:

$$I = \int_0^\infty \frac{dx}{1 + e^{ax}}$$

Difficulty level   ○ Easy   ○ Normal   ● Hard
Calculation amount   ○ Small   ● Normal   ○ Large

1) $\dfrac{1}{a}$

2) $a \ln 2$

3) $\dfrac{1}{a} \ln 2$

4) ∞

9.49. Solve the indefinite integral below.

$$\int \ln x\,dx$$

Difficulty level   ○ Easy   ○ Normal   ● Hard
Calculation amount   ○ Small   ● Normal   ○ Large
1) $x \ln x - x + c$
2) $x \ln x + x + c$
3) $-x \ln x + x + c$
4) $-x \ln x - x + c$

9.50. Solve the following indefinite integral.

$$\int \frac{1}{\sin(x) \cos(x)} \, dx$$

Difficulty level   ○ Easy   ○ Normal   ● Hard
Calculation amount   ○ Small   ● Normal   ○ Large
1) $\ln|\sin(2x)| + c$
2) $\ln|\tan(x)| + c$
3) $\ln|\cos(2x)| + c$
4) $\ln|\cot(x)| + c$

9.51. Calculate the value of the definite integral below.

$$\int_0^{\frac{\pi}{4}} \frac{1}{\cos^4(x)} \, dx$$

Difficulty level   ○ Easy   ○ Normal   ● Hard
Calculation amount   ○ Small   ● Normal   ○ Large
1) $\frac{1}{3}$
2) $\frac{2}{3}$
3) 1
4) $\frac{4}{3}$

9.52. Calculate the value of the definite integral below.

$$\int_0^{\frac{\pi}{4}} \frac{1}{\sqrt[3]{\sin^2(x) \cos^4(x)}} \, dx$$

Difficulty level   ○ Easy   ○ Normal   ● Hard
Calculation amount   ○ Small   ● Normal   ○ Large
1) 1
2) 2
3) 3
4) 4

9.53. Calculate the value of $f(x = e)$ if the derivative of $f(x^2)$ with respect to x is $\frac{6}{x}$ and $f(x = 1) = 0$.
Difficulty level   ○ Easy   ○ Normal   ● Hard
Calculation amount   ○ Small   ● Normal   ○ Large
1) 0
2) 1

3) 3
4) 6

9.54. Calculate the value of $f(x = -1)$ if $f'(cos^2(x)) = \cos(2x)$ and $f(x = 1) = 1$.

Difficulty level   ○ Easy   ○ Normal   ● Hard
Calculation amount   ○ Small   ● Normal   ○ Large
1) 1
2) 2
3) 3
4) 4

9.55. Solve the following indefinite integral.

$$\int \frac{\cos(2x)}{\sin^2(x)\cos^2(x)}\,dx$$

Difficulty level   ○ Easy   ○ Normal   ● Hard
Calculation amount   ○ Small   ● Normal   ○ Large

1) $-\dfrac{2}{\sin(2x)} + c$

2) $-\dfrac{1}{\sin(2x)} + c$

3) $\dfrac{2}{\sin(2x)} + c$

4) $\dfrac{1}{\sin(2x)} + c$

9.56. Calculate the value of the definite integral below.

$$\int_1^e (2x + \ln(x))\,dx$$

Difficulty level   ○ Easy   ○ Normal   ● Hard
Calculation amount   ○ Small   ● Normal   ○ Large
1) e^2
2) 1
3) $1 + e$
4) $e - 1$

9.57. Solve the indefinite integral below.

$$\int \sin(2x)\left(2 + \cos^2(x)\right)^{50}\,dx$$

Difficulty level   ○ Easy   ○ Normal   ● Hard
Calculation amount   ○ Small   ● Normal   ○ Large

1) $\dfrac{1}{51}\left(2 + \cos^2(x)\right)^{51} + c$

2) $-\dfrac{1}{51}\left(2 + \cos^2(x)\right)^{51} + c$

3) $\dfrac{1}{51}\left(2 + \cos^2(x)\right)^{50} + c$

4) $-\dfrac{1}{51}\left(2 + \cos^2(x)\right)^{50} + c$

9.58. Calculate the value of the definite integral below.

$$\int_1^e \frac{\ln(x)}{x}\,dx$$

Difficulty level   ○ Easy   ○ Normal   ● Hard
Calculation amount   ○ Small   ● Normal   ○ Large
1) 1

2) $\dfrac{1}{2}$

3) 2

4) $\dfrac{1}{e}$

9.59. Determine the function of a curve that passes from the point of (0, 1) and the relation below holds.

$$y' = -\frac{2x+2}{4y+1}$$

Difficulty level   ○ Easy   ○ Normal   ● Hard
Calculation amount   ○ Small   ● Normal   ○ Large
1) $2x^2 + y^2 = 34 + 3x$
2) $x^2 - y^2 = -7y + 5$
3) $x^2 + y^2 = 4x + 4y - 1$
4) $x^2 + 2y^2 = -y - 2x + 3$

9.60. Calculate the value of the definite integral below.

$$\int_{\frac{\pi}{6}}^{\frac{\pi}{2}} \frac{\cot(x)}{\sqrt{1 - \cos(2x)}}\,dx$$

Difficulty level   ○ Easy   ○ Normal   ● Hard
Calculation amount   ○ Small   ● Normal   ○ Large
1) $\sqrt{2}$

2) $\dfrac{\sqrt{2}}{2}$

3) $\sqrt{3}$

4) $\dfrac{\sqrt{3}}{2}$

9.61. Calculate the value of the definite integral below.

$$\int_3^6 \frac{x+2}{\sqrt{x-2}}\,dx$$

Difficulty level   ○ Easy   ○ Normal   ● Hard
Calculation amount   ○ Small   ● Normal   ○ Large
1) $\dfrac{25}{3}$

2) $\dfrac{38}{3}$

3) $\dfrac{23}{3}$

4) $\dfrac{34}{3}$

9.62. Calculate the value of the definite integral below.

$$\int_1^4 \frac{\sqrt{1+\sqrt{x}}}{\sqrt{x}}\,dx$$

Difficulty level   ○ Easy   ○ Normal   ● Hard
Calculation amount   ○ Small   ● Normal   ○ Large

1) $2\left(\sqrt{3}+\dfrac{1}{3}\right)$

2) $2\left(\sqrt{3}-\dfrac{1}{3}\right)$

3) $4\left(\sqrt{3}+\dfrac{2\sqrt{2}}{3}\right)$

4) $4\left(\sqrt{3}-\dfrac{2\sqrt{2}}{3}\right)$

9.63. Solve the following indefinite integral if we know that $0 < x < \pi$.

$$\int \cot(x)\sqrt{\sin(x)}\,dx$$

Difficulty level   ○ Easy   ○ Normal   ● Hard
Calculation amount   ○ Small   ● Normal   ○ Large

1) $\sqrt{\cot(x)}+c$
2) $2\sqrt{\sin(x)}+c$
3) $\sin(x)\sqrt{\sin(x)}+c$
4) $\dfrac{1}{2}\sqrt{\sin(x)}+c$

9.64. Calculate the value of the definite integral below.

$$\int_0^{\frac{\pi}{3}} \sec(x)\tan(x)\,dx$$

Difficulty level   ○ Easy   ○ Normal   ● Hard
Calculation amount   ○ Small   ● Normal   ○ Large

1) 1
2) 2
3) $\dfrac{1}{2}$
4) $\dfrac{3}{2}$

9.65. Calculate the value of the definite integral below.

$$\int_{\frac{\pi}{6}}^{\frac{\pi}{4}} \csc(x) \cot(x) dx$$

Difficulty level   ○ Easy   ○ Normal   ● Hard
Calculation amount   ○ Small   ● Normal   ○ Large

1) $2 + \sqrt{2}$
2) $2 - \sqrt{2}$
3) $\sqrt{3} - \sqrt{2}$
4) $\sqrt{3} + \sqrt{2}$

9.66. Calculate the value of the definite integral below.

$$\int_{\frac{\pi}{6}}^{\frac{\pi}{4}} \frac{1}{\sin^2(x) \cos^2(x)} dx$$

Difficulty level   ○ Easy   ○ Normal   ● Hard
Calculation amount   ○ Small   ○ Normal   ● Large

1) $\dfrac{\sqrt{3}}{3}$

2) $\dfrac{2\sqrt{3}}{3}$

3) $\sqrt{3}$
4) $2\sqrt{3}$

9.67. Solve the following indefinite integral.

$$\int (\tan(x) - \cot(x))(\tan(x) + \cot(x))^5 dx$$

Difficulty level   ○ Easy   ○ Normal   ● Hard
Calculation amount   ○ Small   ○ Normal   ● Large

1) $\dfrac{1}{4} (\tan(x) + \cot(x))^4 + c$

2) $\dfrac{1}{5} (\tan(x) + \cot(x))^5 + c$

3) $\dfrac{1}{3} (\tan(x) + \cot(x))^3 + c$

4) $\dfrac{1}{5} (\tan(x) - \cot(x))^5 + c$

9.68. Which one of the choices is not an acceptable solution for the indefinite integral of $\int \sin(x) \cos(x) dx$?

Difficulty level   ○ Easy   ○ Normal   ● Hard
Calculation amount   ○ Small   ○ Normal   ● Large

1) $-\dfrac{1}{4} \cos(2x) + c$

2) $-\dfrac{1}{4} \sin(2x) + c$

3) $-\dfrac{1}{2} \cos^2(x) + c$

4) $\dfrac{1}{2} \sin^2(x) + c$

9.69. Calculate the value of the definite integral below.

$$I = \int_{\ln 2}^{\ln 3} \frac{1 - e^{-2x}}{1 + e^{-2x}} dx$$

Difficulty level ○ Easy ○ Normal ● Hard
Calculation amount ○ Small ○ Normal ● Large

1) $\ln 3 - \ln 2$

2) $3 \ln 2 - \frac{1}{2} \ln 3$

3) $2 \ln 2 - \ln 3$

4) $3 \ln 2 - 2 \ln 3$

References

1. Rahmani-Andebili, M. (2021). Calculus – Practice Problems, Methods, and Solutions, Springer Nature, 2021.
2. Rahmani-Andebili, M. (2021). Precalculus – Practice Problems, Methods, and Solutions, Springer Nature, 2021.

Abstract

In this chapter, the problems of the ninth chapter are fully solved, in detail, step-by-step, and with different methods.

10.1. From list of integral of different functions, we know that [1, 2]:

$$\int u^n(x)du = \frac{1}{n+1}u^{n+1}(x) + c$$

The problem can be solved by defining a new variable as follows:

$$u(x) = 3x + 5$$

$$\Rightarrow du = 3dx \Rightarrow dx = \frac{du}{3}$$

Therefore:

$$I = \int (3x+5)^{17}dx = \int u^{17}(x)\left(\frac{du}{3}\right) = \frac{1}{3}\int u^{17}(x)du$$

$$\Rightarrow I = \frac{1}{3} \times \frac{u^{18}(x)}{18} + c$$

$$\Rightarrow I = \frac{(3x+5)^{18}}{54} + c$$

Choice (2) is the answer.

10.2. From list of integral of different functions, we know that:

$$\int \cos u(x)du = \sin u(x) + c$$

The problem can be solved by defining a new variable as follows:

$$u(x) = 1 + \pi x$$

© The Author(s), under exclusive license to Springer Nature Switzerland AG 2023
M. Rahmani-Andebili, *Calculus I*, https://doi.org/10.1007/978-3-031-45028-0_10

$$\Rightarrow du = \pi dx \Rightarrow dx = \frac{du}{\pi}$$

Thus:

$$I = \int \cos(1 + \pi x) dx = \int (\cos u(x)) \frac{du}{\pi} = \frac{1}{\pi} \int \cos u(x) du$$

$$\Rightarrow I = \frac{1}{\pi} \sin u(x) + c$$

$$\Rightarrow I = \frac{1}{\pi} \sin(1 + \pi x) + c$$

Choice (4) is the answer.

10.3. From the problem, we have:

$$F(x) = \int f(x) dx$$

The problem can be solved by defining a new variable as follows:

$$ax + b = u(x)$$

$$\Rightarrow adx = du \Rightarrow dx = \frac{du}{a}$$

Therefore:

$$I = \int f(ax + b) dx = \frac{1}{a} \int f(u) du$$

$$\Rightarrow I = \frac{F(u)}{a}$$

$$\Rightarrow I = \frac{1}{a} F(ax + b)$$

Choice (4) is the answer.

10.4. From list of integral of functions, we know that:

$$\int x^n dx = \frac{1}{n + 1} x^{n+1} + c$$

The problem can be solved as follows:

$$\int_1^2 \frac{x + 4}{x^3} dx = \int_1^2 \left(\frac{1}{x^2} + \frac{4}{x^3} \right) dx = \left(-\frac{1}{x} - \frac{2}{x^2} \right) \Big|_1^2 = -\frac{1}{2} - \frac{2}{4} - (-1 - 2) = 2$$

Choice (2) is the answer.

10.5. From list of integral of functions, we know that:

$$\int e^u du = e^u$$

The problem can be solved as follows:

$$\int \left(e^x + 2xe^{x^2} \right) dx = e^x + e^{x^2} + c$$

Choice (4) is the answer.

10.6. Based on the information given in the problem, we have:

$$f(x) = \int (x^3 + 5x) dx$$

$$\xrightarrow{\dfrac{d}{dx}} f'(x) = x^3 + 5x \xrightarrow{\dfrac{d}{dx}} f''(x) = 3x^2 + 5$$

$$\xrightarrow{x=1} f''(1) = 3 + 5 = 8$$

Choice (4) is the answer.

10.7. The problem can be solved as follows:

$$f(x) = \int \cos^3(x) dx \Rightarrow f'(x) = \cos^3(x) \Rightarrow f''(x) = -3\cos^2(x)\sin(x)$$

$$f''\left(\frac{\pi}{2}\right) = -3 \times 0 \times 1 = 0$$

Choice (1) is the answer.

10.8. From list of integral of functions, we know that:

$$\int x^n dx = \frac{1}{n+1} x^{n+1} + c$$

The problem can be solved as follows:

$$\int_{-2}^{-1} \frac{x^3 + x^2 - 1}{x^2} dx = \int_{-2}^{-1} \left(x + 1 - \frac{1}{x^2} \right) dx = \left(\frac{x^2}{2} + x + \frac{1}{x} \right) \Big|_{-2}^{-1}$$

$$= \left(\frac{1}{2} - 1 - 1 \right) - \left(2 - 2 - \frac{1}{2} \right) = -1$$

Choice (2) is the answer.

10.9. From list of integral of functions, we know that:

$$\int x^n dx = \frac{1}{n+1} x^{n+1} + c$$

The problem can be solved as follows:

$$\int \frac{x-2}{\sqrt{x}} dx = \int \left(x^{\frac{1}{2}} - 2x^{-\frac{1}{2}} \right) dx = \frac{2}{3} x^{\frac{3}{2}} - 4x^{\frac{1}{2}} + c = \frac{2}{3}\sqrt{x}(x - 6) + c$$

Choice (2) is the answer.

10.10. The problem can be solved by changing the variable of the integral as follows:

$$x^3 + 1 \triangleq u \xrightarrow{\frac{d}{dx}} 3x^2 dx = du \Rightarrow x^2 dx = \frac{du}{3}$$

$$I = \int_0^1 \frac{x^2}{(x^3 + 1)^4} dx = \int_{u_1}^{u_2} u^{-4} \frac{du}{3} = \frac{1}{3} \frac{u^{-3}}{-3} \bigg|_{u_1}^{u_2}$$

$$\Rightarrow I = -\frac{1}{9} (x^3 + 1)^{-3} \bigg|_0^1$$

$$\Rightarrow I = -\frac{1}{9}(1 + 1)^{-3} + \frac{1}{9}(0 + 1)^{-3} = -\frac{1}{72} + \frac{1}{9} = \frac{-1 + 8}{72}$$

$$\Rightarrow I = \frac{7}{72}$$

Choice (4) is the answer.

10.11. From list of integral of different functions, we know that:

$$\int \frac{du}{\sqrt{u(x)}} = 2\sqrt{u(x)}$$

The problem can be solved by defining a new variable as follows:

$$1 + \tan x = u(x)$$

$$\Rightarrow (1 + \tan^2 x) dx = du \Rightarrow \frac{dx}{\cos^2 x} = du$$

Therefore:

$$I = \int \frac{dx}{\cos^2 x \sqrt{1 + \tan x}}$$

$$\Rightarrow I = \int \frac{du}{\sqrt{u}} = 2u^{\frac{1}{2}} + c$$

$$\Rightarrow I = 2\sqrt{1 + \tan x} + c$$

Choice (3) is the answer.

10.12. From list of integral of different functions, we know that:

$$\int \frac{du}{\sqrt{u(x)}} = 2\sqrt{u(x)}$$

The problem can be solved by defining a new variable as follows:

$$u(x) = \ln x$$

$$\Rightarrow du = \frac{dx}{x}$$

Therefore:

$$I = \int \frac{dx}{x\sqrt{\ln x}}$$

$$\Rightarrow I = \int \frac{du}{\sqrt{u(x)}} = 2\sqrt{u(x)} + c$$

$$\Rightarrow I = 2\sqrt{\ln x} + c$$

Choice (3) is the answer.

10.13. From list of integral of functions, we know that:

$$\int \frac{1}{x^2 + a^2} dx = \frac{1}{a} \operatorname{arc}\left(\tan \frac{x}{a}\right) + c$$

Therefore:

$$\int_{-\infty}^{+\infty} \frac{1}{x^2 + 4} dx = \frac{1}{2} \operatorname{arc}\left(\tan\left(\frac{x}{2}\right)\right)\Big|_{-\infty}^{+\infty} = \frac{1}{2}\left(\frac{\pi}{2} - \left(-\frac{\pi}{2}\right)\right) = \frac{\pi}{2}$$

Choice (1) is the answer.

10.14. From list of integral of functions, we know that:

$$\int x^n dx = \frac{1}{n+1} x^{n+1} + c$$

The problem can be solved as follows:

$$\int_1^4 \frac{x - \sqrt{x}}{\sqrt{x}} dx = \int_1^4 \left(x^{\frac{1}{2}} - 1\right) dx = \left(\frac{2}{3} x^{\frac{3}{2}} - x\right)\Big|_1^4 = \left(\frac{16}{3} - 4\right) - \left(\frac{2}{3} - 1\right) = \frac{5}{3}$$

Choice (4) is the answer.

10.15. Based on the information given in the problem, we have:

$$\int f(x)dx = \frac{x^3}{6} \xrightarrow{\frac{d}{dx}} f(x) = \frac{x^2}{2}$$

Therefore:

$$\Rightarrow f\left(\frac{1}{x}\right) = \frac{\left(\frac{1}{x}\right)^2}{2} = \frac{1}{2x^2}$$

$$\Rightarrow \frac{d}{dx}\left(f\left(\frac{1}{x}\right)\right) = \frac{d}{dx}\left(\frac{1}{2x^2}\right) = \frac{-4x}{4x^4} = \frac{-1}{x^3}$$

Choice (1) is the answer.

10.16. As we know:

$$F(x) = \int_{v(x)}^{u(x)} f(x)dx \Rightarrow F'(x) = u'(x)F(u(x)) - v'(x)F(v(x))$$

The problem can be solved as follows:

$$F(\lambda) = \int_0^\lambda \frac{1}{x^4 + 2}dx \Rightarrow F'(\lambda) = 1 \times \frac{1}{\lambda^4 + 2} - 0 = \frac{1}{\lambda^4 + 2} \Rightarrow F'(0) = \frac{1}{2}$$

Choice (2) is the answer.

10.17. Since the function is an odd function and the range of the integral is symmetric, the final answer is zero.

$$\int_{-1}^{1} \frac{x^2}{1 + x^2}\arc(\tan(x))dx = 0$$

Choice (1) is the answer.

10.18. From list of integral of functions, we know that:

$$\int \frac{1}{\sqrt{1 - x^2}}dx = \arc(\sin(x))$$

Therefore:

$$\int_{-\frac{1}{2}}^{\frac{1}{2}} \frac{1}{\sqrt{1 - x^2}}dx = \arc(\sin(x))\Big|_{-\frac{1}{2}}^{\frac{1}{2}} = \frac{\pi}{6} - \left(-\frac{\pi}{6}\right) = \frac{\pi}{3}$$

Choice (2) is the answer.

10.19. From list of integral of functions, we know that:

$$\int \frac{1}{\sqrt{1-u^2}}\,dx = \text{arc}(\sin(u))$$

The problem can be solved by changing the variable of the integral as follows:

$$x - 1 \triangleq u \xrightarrow{\frac{d}{dx}} dx = du$$

$$\int_0^1 \frac{1}{\sqrt{2x-x^2}}\,dx = \int_0^1 \frac{1}{\sqrt{1-(x-1)^2}}\,dx = \int_{u_1}^{u_2} \frac{1}{\sqrt{1-u^2}}\,du = (\text{arc}(\sin(u)))\Big|_{u_1}^{u_2}$$

$$= (\text{arc}(\sin(x-1)))\Big|_0^1 = 0 - \left(-\frac{\pi}{2}\right) = \frac{\pi}{2}$$

Choice (3) is the answer.

10.20. The final answer is zero, since the function is an odd function and the range of the integral is symmetric.

$$\int_{-1}^1 (x^2+1)(x^3+3x)\,dx = 0$$

Choice (2) is the answer.

10.21. From trigonometry, we know that:

$$1 + \cos(u) = 2\cos^2\left(\frac{u}{2}\right)$$

$$\frac{1}{\cos^2\left(\frac{u}{2}\right)} = 1 + \tan^2\left(\frac{u}{2}\right)$$

In addition, from list of integral of functions, we know that:

$$\int \left(1 + \tan^2\left(\frac{u}{a}\right)\right)du = a\tan\left(\frac{u}{a}\right) + c$$

The problem can be solved by changing the variable of the integral as follows:

$$\cos(x) \triangleq u \Rightarrow \frac{d}{dx}\cos(x) = \frac{d}{dx}u \Rightarrow -\sin(x)dx = du$$

$$\Rightarrow \int \frac{\sin(x)}{1+\cos(\cos(x))}\,dx = -\int \frac{1}{1+\cos(u)}\,du = -\int \frac{1}{2\cos^2\left(\frac{u}{2}\right)}\,du$$

$$= -\frac{1}{2}\int \left(1 + \tan^2\left(\frac{u}{2}\right)\right)du = -\tan\left(\frac{u}{2}\right) + c = -\tan\left(\frac{\cos(x)}{2}\right) + c$$

Choice (1) is the answer.

10.22. From list of integral of functions, we know that:

$$\int x^n dx = \frac{1}{n+1}x^{n+1} + c$$

Based on the information given in the problem, we have:

$$y(3) = 4$$

$$y' = -\frac{x}{y}$$

The problem can be solved as follows:

$$y' = -\frac{x}{y} \Rightarrow yy' + x = 0 \xrightarrow{\int dx} \frac{y^2}{2} + \frac{x^2}{2} = c' \Rightarrow y^2 + x^2 = c \qquad (1)$$

$$\xrightarrow{y(3) = 4} 4^2 + 3^2 = c \Rightarrow c = 25 \qquad (2)$$

$$\xrightarrow{(1), (2)} y^2 + x^2 = 25$$

Choice (4) is the answer.

10.23. From list of integral of functions, we know that:

$$\int x^n dx = \frac{1}{n+1}x^{n+1} + c$$

The problem can be solved as follows:

$$\int \frac{x}{\sqrt{x-1}}dx = \int \frac{x-1+1}{\sqrt{x-1}}dx = \int \left((x-1)^{\frac{1}{2}} + (x-1)^{-\frac{1}{2}}\right)dx = \frac{2}{3}(x-1)^{\frac{3}{2}} + 2(x-1)^{\frac{1}{2}} + c$$

Choice (2) is the answer.

10.24. From list of integral of functions, we know that:

$$\int x^n dx = \frac{1}{n+1}x^{n+1} + c$$

The problem can be solved as follows:

$$\int_{-2}^{5} |x - 3| dx = \int_{-2}^{3}(3-x)dx + \int_{3}^{5}(x-3)dx = \left(3x - \frac{x^2}{2}\right)\Big|_{-2}^{3} + \left(\frac{x^2}{2} - 3x\right)\Big|_{3}^{5}$$

$$= \left(9 - \frac{9}{2}\right) - (-6-2) + \left(\frac{25}{2} - 15\right) - \left(\frac{9}{2} - 9\right) = \frac{9}{2} + 8 - \frac{5}{2} + \frac{9}{2} = \frac{9 + 16 - 5 + 9}{2} = \frac{29}{2}$$

Choice (3) is the answer.

10.25. From list of integral of functions, we know that:

$$\int x^n dx = \frac{1}{n+1}x^{n+1} + c$$

The problem can be solved as follows:

$$\int_1^2 \frac{x(x+1)^2 + 2}{(x+1)^2} dx = \int_1^2 \left(x + 2(x+1)^{-2}\right) dx = \left(\frac{x^2}{2} - \frac{2}{x+1}\right)\Big|_1^2 = \left(2 - \frac{2}{3}\right) - \left(\frac{1}{2} - 1\right) = \frac{4}{3} + \frac{1}{2} = \frac{11}{6}$$

Choice (3) is the answer.

10.26. From list of integral of functions, we know that:

$$\int x^n dx = \frac{1}{n+1}x^{n+1} + c$$

$$\int \frac{1}{u} du = \ln|u| + c$$

The problem can be solved as follows:

$$\int \frac{1}{1+e^x} dx = \int \frac{1 + e^x - e^x}{1 + e^x} dx = \int \left(1 - \frac{e^x}{1+e^x}\right) dx = \int 1 dx - \int \frac{e^x}{1+e^x} dx$$

$$= x + c' - \int \frac{e^x}{1+e^x} dx \qquad (1)$$

Now, we should change the variable of the integral as follows:

$$1 + e^x \triangleq u \Rightarrow e^x dx = du \qquad (2)$$

Solving (1) and (2):

$$x + c' - \int \frac{1}{u} du = x + c' - \ln|u| + c'' = x - \ln|1 + e^x| + c = x - \ln(1 + e^x) + c$$

Choice (2) is the answer.

10.27. The problem can be solved as follows:

$$y' = \frac{x+1}{1-y} \Rightarrow y' - yy' = x + 1 \xrightarrow{\int dx} y - \frac{y^2}{2} = \frac{x^2}{2} + x + c$$

$$\xrightarrow{(x,y) = (1,1)} 1 - \frac{1}{2} = \frac{1}{2} + 1 + c \Rightarrow c = -1$$

$$\Rightarrow y - \frac{y^2}{2} = \frac{x^2}{2} + x - 1 \Rightarrow x^2 + y^2 + 2x - 2y - 2 = 0$$

Choice (1) is the answer.

10.28. The problem can be solved as follows:

$$y' = \frac{3x}{2y} \Rightarrow 2yy' = 3x \Rightarrow y^2 = \frac{3}{2}x^2 + c$$

$$\xrightarrow{(x, y) = (1, 1)} 1 = \frac{3}{2} + c \Rightarrow c = \frac{-1}{2}$$

$$\Rightarrow y^2 = \frac{3}{2}x^2 - \frac{1}{2} \Rightarrow 2y^2 - 3x^2 + 1 = 0$$

Choice (1) is the answer.

10.29. Based on the information given in the problem, we have:

$$\int \frac{3x}{\sqrt{x^2 + 1}} dx = A\sqrt{x^2 + 1} + c \qquad\qquad (1)$$

From list of integral of functions, we know that:

$$\int \frac{1}{\sqrt{u}} du = 2\sqrt{u} + c$$

The problem can be solved by changing the variable of the integral as follows:

$$x^2 + 1 \triangleq u \xrightarrow{\frac{d}{dx}} 2xdx = du \Rightarrow xdx = \frac{1}{2} du$$

$$\int \frac{3x}{\sqrt{x^2 + 1}} dx = \int \frac{3 \times \frac{1}{2} du}{\sqrt{u}} = \frac{3}{2} \int \frac{1}{\sqrt{u}} du = \frac{3}{2} \times 2\sqrt{u} = 3\sqrt{x^2 + 1} + c \qquad (2)$$

Therefore:

$$\xrightarrow{(1), (2)} A\sqrt{x^2 + 1} + c = 3\sqrt{x^2 + 1} + c \Rightarrow A = 3$$

Choice (4) is the answer.

10.30. From list of integral of functions, we know that:

$$\int x^n dx = \frac{1}{n + 1} x^{n+1} + c$$

The problem can be solved as follows:

$$\int_{-1}^{2} |x|dx = \int_{-1}^{0} |x|dx + \int_{0}^{2} |x|dx = \int_{-1}^{0} (-x)dx + \int_{0}^{2} xdx = -\frac{x^2}{2}\Big|_{-1}^{0} + \frac{x^2}{2}\Big|_{0}^{2}$$

$$= -\left(0 - \frac{1}{2}\right) + (2 - 0) = \frac{5}{2}$$

Choice (2) is the answer.

Note that for this problem, a graphical method can be used which is faster than the abovementioned method. Think about it!

10.31. From trigonometry, we know that:

$$1 - \cos(2x) = 2\sin^2(x)$$

Furthermore, from list of integral of functions, we know that:

$$\int x^n dx = \frac{1}{n+1}x^{n+1} + c$$

$$\int \cos(ax)dx = \frac{1}{a}\sin(ax) + c$$

The problem can be solved as follows:

$$\int_0^{\frac{\pi}{2}} \sin^2(x)dx = \int_0^{\frac{\pi}{2}} \left(\frac{1}{2} - \frac{\cos(2x)}{2}\right)dx = \left(\frac{1}{2}x - \frac{1}{4}\sin(2x)\right)\Big|_0^{\frac{\pi}{2}} = \left(\frac{1}{2} \times \frac{\pi}{2} - 0\right) - (0) = \frac{\pi}{4}$$

Choice (3) is the answer.

10.32. From list of integral of functions, we know that:

$$\int x^n dx = \frac{1}{n+1}x^{n+1} + c$$

The problem can be solved by changing the variable of the integral as follows:

$$\tan(x) \triangleq u \xrightarrow{\frac{d}{dx}} \left(1 + \tan^2(x)\right)dx = du$$

$$\int_0^{\frac{3\pi}{4}} \left(\tan^5(x) + \tan^7(x)\right)dx = \int_0^{\frac{3\pi}{4}} \tan^5(x)\left(1 + \tan^2(x)\right)dx$$

$$= \int_{u_1}^{u_2} u^5 du = \frac{1}{6}u^6\Big|_{u_1}^{u_2} = \frac{1}{6}\tan^6(x)\Big|_0^{\frac{3\pi}{4}} = \frac{1}{6}(-1)^6 - 0 = \frac{1}{6}$$

Choice (4) is the answer.

10.33. The problem can be solved as follows:

$$y' = y^2\cos(x) \Rightarrow \frac{y'}{y^2} = \cos(x) \Rightarrow \frac{-1}{y} = \sin(x) + c$$

$$\xrightarrow{(x,y) = (\pi, 1)} -1 = 0 + c \Rightarrow c = -1 \Rightarrow \frac{-1}{y} = \sin(x) - 1$$

We need to check each choice as follows:

$$\text{Choice 1}: \xrightarrow{(x,y) = \left(\frac{3\pi}{2}, 2\right)} \frac{-1}{2} = \sin\left(\frac{3\pi}{2}\right) - 1 \Rightarrow \frac{-1}{2} \neq -2$$

$$\text{Choice 2}: \xrightarrow{(x, y) = \left(\frac{\pi}{2}, -1\right)} \frac{-1}{-1} = \sin\left(\frac{\pi}{2}\right) - 1 \Rightarrow 1 \neq 0$$

$$\text{Choice 3}: \xrightarrow{(x, y) = \left(\frac{\pi}{2}, 1\right)} \frac{-1}{1} = \sin\left(\frac{\pi}{2}\right) - 1 \Rightarrow -1 \neq 0$$

$$\text{Choice 4}: \xrightarrow{(x, y) = (0, 1)} \frac{-1}{1} = \sin(0) - 1 \Rightarrow -1 = -1$$

Choice (4) is the answer.

10.34. From list of integral of functions, we know that:

$$\int x^n dx = \frac{1}{n+1} x^{n+1} + c$$

Based on the information given in the problem, we have:

$$I_2 = \int_3^5 \frac{1}{x-2} dx = m \tag{1}$$

The problem can be solved as follows:

$$I_1 = \int_3^5 \frac{3x}{x-2} dx = \int_3^5 \frac{3x - 6 + 6}{x-2} dx = \int_3^5 3 dx + 6\int_3^5 \frac{1}{x-2} dx \tag{2}$$

Solving (1) and (2):

$$I_1 = \int_3^5 \frac{3x}{x-2} dx = 3x\Big|_3^5 + 6m = 15 - 9 + 6m = 6 + 6m$$

Choice (3) is the answer.

10.35. From list of integral of functions, we know that:

$$\int \frac{1}{u} du = \ln|u| + c$$

The problem can be solved as follows:

$$\int_2^3 \frac{x}{x^2 - 1} dx = \frac{1}{2}\int_2^3 \frac{2x}{x^2 - 1} dx \tag{1}$$

Now, we need to change the variable of the integral as follows:

$$x^2 - 1 \triangleq u \xrightarrow{\frac{d}{dx}} 2x dx = du \tag{2}$$

Solving (1) and (2):

$$\frac{1}{2} \int_{u_1}^{u_2} \frac{1}{u} \, du = \frac{1}{2} \ln|u| \Big|_{u_1}^{u_2} = \frac{1}{2} \ln|x^2 - 1| \Big|_{2}^{3} = \frac{1}{2} \ln 8 - \frac{1}{2} \ln 3 = \frac{1}{2} \ln \frac{8}{3} = \ln \sqrt{\frac{8}{3}}$$

Choice (4) is the answer.

10.36. From list of integral of functions, we know that:

$$\int x^n \, dx = \frac{1}{n+1} x^{n+1} + c$$

As we know:

$$F(x) = \int_{v(x)}^{u(x)} f(x) \, dx \Rightarrow F'(x) = u'(x)F(u(x)) - v'(x)F(v(x))$$

Therefore:

$$f(x) = \int_{a}^{x} \sqrt{t} \, dt \Rightarrow f'(x) = \sqrt{x}$$

$$\int_{0}^{4} f'(x) \, dx = \int_{0}^{4} x^{\frac{1}{2}} \, dx = \left(\frac{2}{3} x^{\frac{3}{2}} \right) \Big|_{0}^{4} = \frac{16}{3}$$

Choice (2) is the answer.

10.37. From list of integral of functions, we know that:

$$\int x^n \, dx = \frac{1}{n+1} x^{n+1} + c$$

The problem can be solved by changing the variable of the integral as follows:

$$\tan(x) \triangleq u \xrightarrow{\frac{d}{dx}} (1 + \tan^2(x)) \, dx = du$$

$$\int 8(\tan^6(x) + \tan^8(x)) \, dx = 8 \int \tan^6(x)(1 + \tan^2(x)) \, dx$$

$$= 8 \int u^6 \, du = \frac{8}{7} u^7 + c = \frac{8}{7} \tan^7(x) + c$$

Choice (4) is the answer.

10.38. From list of integral of functions, we know that:

$$\int x^n \, dx = \frac{1}{n+1} x^{n+1} + c$$

The problem can be solved as follows:

$$\int_{-1}^{1} (x+1)(x^2+2x+3)dx = \int_{-1}^{1} (x+1)\Big((x+1)^2+2\Big)dx = \int_{-1}^{1} \Big((x+1)^3+2(x+1)\Big)dx \tag{1}$$

Now, we should change the variable of the integral as follows:

$$x+1 \triangleq u \Rightarrow dx = du \tag{2}$$

Solving (1) and (2):

$$\int_{-1}^{1} (u^3+2u)dx = \frac{u^4}{4}+u^2$$

$$= \left(\frac{(x+1)^4}{4}+(x+1)^2\right)\Big|_{-1}^{1} = \frac{16}{4}+4-0 = 8$$

Choice (3) is the answer.

10.39. From list of integral of functions, we know that:

$$\int x^n dx = \frac{1}{n+1}x^{n+1}+c$$

The problem can be solved as follows:

$$\int_{\frac{1}{2}}^{1} \left[\frac{1}{x}\right] \times \frac{1}{x^3}dx = \int_{\frac{1}{2}}^{1} 1 \times x^{-3}dx = \left(\frac{x^{-2}}{-2}\right)\Big|_{\frac{1}{2}}^{1} = \left(\frac{-1}{2x^2}\right)\Big|_{\frac{1}{2}}^{1} = \frac{-1}{2}-(-2) = \frac{3}{2}$$

Choice (4) is the answer.

10.40. As we know:

$$F(x) = \int_{v(x)}^{u(x)} f(x)dx \Rightarrow F'(x) = u'(x)F(u(x)) - v'(x)F(v(x))$$

Therefore:

$$y'_x = u'_x + v'_x = \left(2x\frac{\sin(x^2)}{x^2}-0\right) + \left(0-2x\frac{\sin(x^2)}{x^2}\right) = 0$$

Choice (3) is the answer.

10.41. Based on the information given in the problem, we know that:

$$h(x) = \int f'(3x+2)dx \tag{1}$$

$$h(0) = 1 \tag{2}$$

$$f(2) = 3 \tag{3}$$

We should change the variable of the integral of $h(x)$ as follows:

$$f(3x + 2) \triangleq u \Rightarrow 3f'(3x + 2)dx = du \tag{4}$$

Solving (1) and (4):

$$h(x) = \int \frac{1}{3} du = \frac{1}{3} u + c = \frac{1}{3} f(3x + 2) + c \tag{5}$$

Solving (2) and (5):

$$1 = \frac{1}{3} f(2) + c \xrightarrow{Using\ (3)} 1 = \frac{1}{3} \times 3 + c \Rightarrow c = 0 \Rightarrow h(x) = \frac{1}{3} f(3x + 2)$$

$$\Rightarrow f(3x + 2) = 3h(x)$$

Choice (3) is the answer.

10.42. From list of integral of functions, we know that:

$$\int x^n du = \frac{1}{n + 1} x^{n+1} + c$$

Based on the information given in the problem, we know that:

$$y''(x) = 2x + 1 \tag{1}$$

The curve is tangent to $y = x$ in the origin. Thus:

$$y'(0) = 1 \tag{2}$$

$$y(0) = 0 \tag{3}$$

Applying integral operation on (1):

$$\xrightarrow{\int dx} y'(x) = x^2 + x + a \tag{4}$$

Applying integral operation on (4):

$$\xrightarrow{\int dx} y(x) = \frac{x^3}{3} + \frac{x^2}{2} + ax + b \tag{5}$$

Solving (2) and (4):

$$a = 1 \tag{6}$$

Solving (3) and (5):

$$0 = 0 + 0 + 0 + b \Rightarrow b = 0 \qquad (7)$$

Solving (5)–(7):

$$y(x) = \frac{x^3}{3} + \frac{x^2}{2} + x$$

Now, we need to check the choices as follows:

$$y(1) = \frac{1}{3} + \frac{1}{2} + 1 = \frac{2 + 3 + 6}{6} = \frac{11}{6}$$

$$y(2) = \frac{2^3}{3} + \frac{2^2}{2} + 2 = \frac{16 + 12 + 12}{6} = \frac{20}{3}$$

Choice (1) is the answer.

10.43. From trigonometry, we know that:

$$1 + \cos(2x) = 2\cos^2(x)$$

In addition, from list of integral of functions, we know that:

$$\int \sin(ax)dx = -\frac{1}{a}\cos(ax) + c$$

The problem can be solved as follows:

$$\int \sin(2x)\cos(4x)dx = \int \sin(2x)\left(2\cos^2(2x) - 1\right)dx$$

$$= \int \cos^2(2x) \times 2\sin(2x)dx - \int \sin(2x)dx \qquad (1)$$

Now, we should change the variable of the integral as follows:

$$\cos(2x) \triangleq u \Rightarrow -2\sin(2x)dx = du \qquad (2)$$

Solving (1) and (2):

$$-\int u^2 du - \int \sin(2x)dx = -\frac{1}{3}u^3 + \frac{1}{2}\cos(2x) + c = -\frac{1}{3}\cos^3(2x) + \frac{1}{2}\cos(2x) + c$$

Choice (2) is the answer.

10.44. From list of integral of different functions, we know that:

$$\int e^{u(x)}du = e^{u(x)} + c$$

The problem can be solved by defining a new variable as follows:

$$\arctan x = u(x)$$

$$\Rightarrow \frac{dx}{1+x^2} = du$$

Therefore:

$$\Rightarrow I = \int \frac{e^{\arctan x}}{1+x^2}\, dx$$

$$\Rightarrow I = \int e^u du = e^u + c$$

$$\Rightarrow I = e^{\arctan x} + c$$

Choice (1) is the answer.

10.45. From list of integral of functions, we know that:

$$\int x^n du = \frac{1}{n+1} x^{n+1} + c$$

The problem can be solved as follows:

$$\int_{-1}^{1} \left[\frac{x}{3}\right] dx = \int_{-1}^{0} (-1)dx + \int_{0}^{1} 0\, dx = (-x)\Big|_{-1}^{0} + 0 = 0 - (1) = -1$$

Choice (3) is the answer.

10.46. The problem can be heuristically solved as follows:

$$xy' + y = 1 \Rightarrow (xy)' = 1 \Rightarrow xy = x + c$$

$$\xrightarrow{(x, y) = (1, 2)} 1 \times 2 = 1 + c \Rightarrow c = 1$$

$$\Rightarrow xy = x + 1 \Rightarrow y = 1 + \frac{1}{x}$$

Choice (1) is the answer.

10.47. The problem can be solved as follows:

$$\int \frac{f'(\sqrt[3]{x})}{\sqrt[3]{x^2}} dx = 3\int \frac{1}{3\sqrt[3]{x^2}} f'(\sqrt[3]{x}) dx \tag{1}$$

Now, we should change the variable of the integral as follows:

$$f(\sqrt[3]{x}) \triangleq u \Rightarrow \frac{1}{3\sqrt[3]{x^2}} f'(\sqrt[3]{x}) dx = du \tag{2}$$

Solving (1) and (2):

$$3\int du = 3u + c = 3f(\sqrt[3]{x}) + c$$

Choice (4) is the answer.

10.48. The problem can be solved as follows:

$$I = \int_0^\infty \frac{dx}{1 + e^{ax}}$$

$$\xrightarrow{\times \frac{e^{-ax}}{e^{-ax}}} I = \int_0^\infty \frac{e^{-ax}}{e^{-ax} + 1} dx$$

From list of integral of different functions, we know that:

$$\int \frac{du}{u(x)} = \ln u(x)$$

The problem can be solved by defining a new variable as follows:

$$e^{-ax} + 1 = u(x)$$

$$\Rightarrow -ae^{-ax}dx = du \Rightarrow e^{-ax}dx = \frac{du}{-a}$$

Therefore:

$$I = \int_0^\infty \frac{e^{-ax}}{e^{-ax} + 1} dx$$

$$\Rightarrow I = -\frac{1}{a} \int_{u_1}^{u_2} \frac{du}{u(x)} = -\frac{1}{a} [\ln u(x)]_{u_1}^{u_2}$$

$$\Rightarrow I = [\ln(e^{-ax} + 1)]_0^\infty$$

$$\Rightarrow I = -\frac{1}{a} [\ln 1 - \ln 2]$$

$$\Rightarrow I = \frac{1}{a} \ln 2$$

Choice (3) is the answer.

10.49. From integration by parts (partial integration), we know that:

$$\int u(x)dv = u(x)v(x) - \int v(x)du$$

In addition, from list of integral of functions, we know that:

$$\int \frac{1}{u} du = \ln|u| + c$$

The problem can be solved as follows:

$$\int \ln x\,dx \Rightarrow \begin{cases} u(x) = \ln x \\ dv = dx \end{cases} \Rightarrow \begin{cases} du = \dfrac{dx}{x} \\ v(x) = x \end{cases}$$

$$\Rightarrow \int \ln x\,dx = x\ln x - \int dx = x\ln x - x + c$$

Choice (1) is the answer.

10.50. From trigonometry, we know that:

$$\frac{1}{\cos^2(x)} = 1 + \tan^2(x)$$

$$\frac{\sin(x)}{\cos(x)} = \tan(x)$$

Moreover, from list of integral of functions, we know that:

$$\int \frac{1}{u}\,du = \ln|u| + c$$

The problem can be solved as follows:

$$\int \frac{1}{\sin(x)\cos(x)}\,dx = \int \frac{1}{\sin(x)\cos(x)\times\frac{\cos(x)}{\cos(x)}}\,dx = \int \frac{1}{\frac{\sin(x)}{\cos(x)}\cos^2(x)}\,dx = \int \frac{1+\tan^2(x)}{\tan(x)}\,dx \qquad (1)$$

Now, we should change the variable of the integral as follows:

$$\tan(x) \triangleq u \Rightarrow (1 + \tan^2(x))dx = du \qquad (2)$$

Solving (1) and (2):

$$\int \frac{1}{u}\,du = \ln|u| + c = \ln|\tan(x)| + c$$

Choice (2) is the answer.

10.51. From trigonometry, we know that:

$$1 + \tan^2(x) = \frac{1}{\cos^2(x)}$$

The problem can be solved as follows:

$$\int_0^{\frac{\pi}{4}} \frac{1}{\cos^4(x)}\,dx = \int_0^{\frac{\pi}{4}} (1 + \tan^2(x))(1 + \tan^2(x))dx \qquad (1)$$

Now, we should change the variable of the integral as follows:

$$\tan(x) \triangleq u \xrightarrow{\frac{d}{dx}} (1 + \tan^2(x))dx = du \qquad (2)$$

Solving (1) and (2):

$$\int_{u_1}^{u_2} \left(1 + u^2\right) du = \left(u + \frac{1}{3} u^3\right)\Big|_{u_1}^{u_2} = \left(\tan(x) + \frac{1}{3} \tan^3(x)\right)\Big|_0^{\frac{\pi}{4}} = \left(1 + \frac{1}{3}\right) - 0 = \frac{4}{3}$$

Choice (4) is the answer.

10.52. From trigonometry, we know that:

$$1 + \tan^2(x) = \frac{1}{\cos^2(x)}$$

$$\tan(x) = \frac{\sin(x)}{\cos(x)}$$

The problem can be solved as follows:

$$\int_{\frac{\pi}{6}}^{\frac{\pi}{4}} \frac{1}{\sqrt[3]{\sin^2(x)\cos^4(x)}} dx = \int_0^{\frac{\pi}{4}} \frac{1}{\sqrt[3]{\sin^2(x)\cos^4(x) \times \frac{\cos^2(x)}{\cos^2(x)}}} dx = \int_0^{\frac{\pi}{4}} \frac{\frac{1}{\cos^2(x)}}{\sqrt[3]{\frac{\sin^2(x)}{\cos^2(x)}}} dx$$

$$= \int_{\frac{\pi}{6}}^{\frac{\pi}{4}} \tan^{\frac{-2}{3}}(x)\left(1 + \tan^2(x)\right) dx \tag{1}$$

Now, we should change the variable of the integral as follows:

$$\tan(x) \triangleq u \xrightarrow{\frac{d}{dx}} \left(1 + \tan^2(x)\right) dx = du \tag{2}$$

Solving (1) and (2):

$$\int_{u_1}^{u_2} u^{-\frac{2}{3}} du = 3u^{\frac{1}{3}}\Big|_{u_1}^{u_2} = \left(3 \tan^{\frac{1}{3}}(x)\right)\Big|_0^{\frac{\pi}{4}} = 3(1 - 0) = 3$$

Choice (3) is the answer.

10.53. From list of integral of functions, we know that:

$$\int \frac{1}{u} du = \ln|u| + c$$

Based on the information given in the problem, we have:

$$f(1) = 0 \tag{1}$$

$$\frac{d}{dx}\left(f\left(x^2\right)\right) = \frac{6}{x} \tag{2}$$

The problem can be solved as follows:

$$\frac{d}{dx}\left(f\left(x^2\right)\right) = 2xf'\left(x^2\right) \tag{3}$$

$$\xrightarrow{(2),(3)} \frac{6}{x} = 2xf'\left(x^2\right) \Rightarrow f'\left(x^2\right) = \frac{3}{x^2} \tag{4}$$

By changing the variable of the integral, we have:

$$x^2 \triangleq t \tag{5}$$

$$\xrightarrow{(4),(5)} f'(t) = \frac{3}{t} \xrightarrow{\int dt} f(t) = 3\ln|t| + c \tag{6}$$

$$\xrightarrow{(1),(6)} 0 = 3 \times 0 + c \Rightarrow c = 0 \Rightarrow f(t) = 3\ln|t|$$

$$\Rightarrow f(e) = 3\ln(e) = 3 \times 1 = 3$$

Choice (3) is the answer.

10.54. From list of integral of functions, we know that:

$$\int x^n dx = \frac{1}{n+1}x^{n+1} + c$$

From trigonometry, we know that:

$$1 + \cos(2x) = 2\cos^2(x)$$

Moreover, based on the information given in the problem, we have:

$$f(1) = 1 \tag{1}$$

$$f'\left(\cos^2(x)\right) = \cos(2x) \tag{2}$$

The problem can be solved as follows:

$$f'\left(\cos^2(x)\right) = \cos(2x) = 2\cos^2(x) - 1 \tag{3}$$

By changing the variable of the integral, we have:

$$\cos^2(x) \triangleq t \tag{4}$$

$$\xrightarrow{(3),(4)} f'(t) = 2t - 1 \xrightarrow{\int dt} f(t) = t^2 - t + c \tag{5}$$

$$\xrightarrow{(1),(5)} 1 = 1 - 1 + c \Rightarrow c = 1 \Rightarrow f(t) = t^2 - t + 1$$

$$\Rightarrow f(-1) = (-1)^2 - (-1) + 1 = 3$$

Choice (3) is the answer.

10.55. From list of integral of functions, we know that:

$$\int x^n dx = \frac{1}{n+1} x^{n+1} + c$$

From trigonometry, we know that:

$$\sin(2x) = 2\sin(x)\cos(x)$$

The problem can be solved as follows:

$$\int \frac{\cos(2x)}{\sin^2(x)\cos^2(x)} dx = \int \frac{\cos(2x)}{\left(\frac{1}{2}\sin(2x)\right)^2} dx = \int \frac{\cos(2x)}{\frac{1}{4}\sin^2(2x)} dx = \int 4\cos(2x)(\sin(2x))^{-2} dx$$

Now, we need to change the variable of the integral as follows:

$$\sin(2x) \triangleq u \xrightarrow{\frac{d}{dx}} 2\cos(2x)dx = du$$

$$\Rightarrow \int 2u^{-2} du = -\frac{2}{u} + c = \frac{-2}{\sin(2x)} + c$$

Choice (1) is the answer.

10.56. From integration by parts (partial integration), we know that:

$$\int \ln(x)dx = x\ln|x| - x$$

or, in general:

$$\int u(x)dv = u(x)v(x) - \int v(x)du$$

In addition, from list of integral of functions, we know that:

$$\int \frac{1}{u} du = \ln|u| + c$$

$$\int x^n dx = \frac{1}{n+1} x^{n+1} + c$$

The problem can be solved as follows:

$$\int_1^e (2x + \ln(x))dx = \int_1^e 2x\, dx + \int_1^e \ln(x)dx = (x^2)\Big|_1^e + (x\ln|x| - x)\Big|_1^e$$

$$= e^2 - 1 + (e - e) - (0 - 1) = e^2$$

Choice (1) is the answer.

10.57. From list of integral of functions, we know that:

$$\int x^n dx = \frac{1}{n+1} x^{n+1} + c$$

The problem can be solved as follows:

$$\int \sin(2x)\left(2 + \cos^2(x)\right)^{50} dx \tag{1}$$

We should change the variable of the integral as follows:

$$2 + \cos^2(x) \triangleq u \Rightarrow -2\cos(x)\sin(x)dx = du \Rightarrow -\sin(2x) = du \tag{2}$$

Solving (1) and (2):

$$-\int u^{50} du = -\frac{u^{51}}{51} + c = -\frac{1}{51}\left(2 + \cos^2(x)\right)^{51} + c$$

Choice (2) is the answer.

10.58. From list of integral of functions, we know that:

$$\int u^n du = \frac{1}{n+1} u^{n+1} + c$$

The problem can be solved as follows:

$$\int_1^e \frac{\ln(x)}{x} dx = \int_1^e \ln(x)\left(\frac{1}{x}\right) dx \tag{1}$$

Now, we should change the variable of the integral as follows:

$$\ln(x) \triangleq u \Rightarrow \frac{1}{x}dx = du \tag{2}$$

Solving (1) and (2):

$$\int_{u_1}^{u_2} u du = \frac{1}{2}u^2 \Big|_{u_1}^{u_2} = \frac{1}{2}\left(\ln(x)\right)^2 \Big|_1^e = \frac{1}{2} - 0 = \frac{1}{2}$$

Choice (2) is the answer.

10.59. From list of integral of functions, we know that:

$$\int u^n du = \frac{1}{n+1} u^{n+1} + c$$

The problem can be solved as follows:

$$y' = -\frac{2x+2}{4y+1} \Rightarrow 4yy' + y' = -2x-2 \xrightarrow{\int dx} 2y^2 + y = -x^2 - 2x + c \tag{1}$$

$$\xrightarrow{(x,y)=(0,1)} 2 + 1 = 0 + c \Rightarrow c = 3 \tag{2}$$

Solving (1) and (2):

$$2y^2 + y = -x^2 - 2x + 3 \Rightarrow x^2 + 2y^2 = -y - 2x + 3$$

Choice (4) is the answer.

10.60. From list of integral of functions, we know that:

$$\int u^n du = \frac{1}{n+1} u^{n+1} + c$$

Moreover, from trigonometry, we know that:

$$\cot(x) = \frac{\cos(x)}{\sin(x)}$$

$$1 - \cos(2x) = 2\sin^2(x)$$

The problem can be solved as follows:

$$\int_{\frac{\pi}{6}}^{\frac{\pi}{2}} \frac{\cot(x)}{\sqrt{1-\cos(2x)}} dx = \int_{\frac{\pi}{6}}^{\frac{\pi}{2}} \frac{\cos(x)}{\sqrt{2\sin^2(x)}\sin(x)} dx = \int_{\frac{\pi}{6}}^{\frac{\pi}{2}} \frac{\cos(x)}{\sqrt{2}|\sin(x)|\sin(x)} dx$$

$$= \frac{\sqrt{2}}{2} \int_{\frac{\pi}{6}}^{\frac{\pi}{2}} (\sin(x))^{-2} \cos(x) dx \tag{1}$$

Now, we should change the variable of the integral as follows:

$$\sin(x) \triangleq u \Rightarrow \cos(x)dx = du \tag{2}$$

Solving (1) and (2):

$$\frac{\sqrt{2}}{2} \int_{u_1}^{u_2} u^{-2} du = -\frac{\sqrt{2}}{2} u^{-1} \Big|_{u_1}^{u_2} = \left(-\frac{\sqrt{2}}{2} \times \frac{1}{\sin(x)} \right) \Big|_{\frac{\pi}{6}}^{\frac{\pi}{2}} = -\frac{\sqrt{2}}{2}(1-2) = \frac{\sqrt{2}}{2}$$

Choice (2) is the answer.

10.61. From list of integral of functions, we know that:

$$\int u^n du = \frac{1}{n+1} u^{n+1} + c$$

The problem can be solved as follows:

$$\int_3^6 \frac{x+2}{\sqrt{x-2}}dx = \int_3^6 \frac{x-2+4}{\sqrt{x-2}}dx = \int_3^6 \left((x-2)^{\frac{1}{2}} + 4(x-2)^{-\frac{1}{2}}\right)dx$$

$$= \left(\frac{2}{3}(x-2)^{\frac{3}{2}} + 4 \times 2(x-2)^{\frac{1}{2}}\right)\Big|_3^6 = \left(\frac{2}{3}(4)^{\frac{3}{2}} + 4 \times 2(4)^{\frac{1}{2}}\right) - \left(\frac{2}{3}(1)^{\frac{3}{2}} + 4 \times 2(1)^{\frac{1}{2}}\right)$$

$$= \left(\frac{16}{3} + 16\right) - \left(\frac{2}{3} + 8\right) = \frac{14}{3} + 8 = \frac{38}{3}$$

Choice (2) is the answer.

10.62. In addition, from list of integral of functions, we know that:

$$\int u^n du = \frac{1}{n+1}u^{n+1} + c$$

The problem can be solved as follows:

$$\int_1^4 \frac{\sqrt{1+\sqrt{x}}}{\sqrt{x}}dx = 2\int_1^4 (1+\sqrt{x})^{\frac{1}{2}} \times \frac{1}{2\sqrt{x}}dx \tag{1}$$

Now, we should change the variable of the integral as follows:

$$1 + \sqrt{x} \triangleq u \Rightarrow \frac{1}{2\sqrt{x}}dx = du \tag{2}$$

Solving (1) and (2):

$$2\int_{u_1}^{u_2} u^{\frac{1}{2}}du = \left(2 \times \frac{2}{3}u^{\frac{3}{2}}\right)\Big|_{u_1}^{u_2} = \left(\frac{4}{3}(1+\sqrt{x})^{\frac{3}{2}}\right)\Big|_1^4 = \frac{4}{3}\left(3\sqrt{3} - 2\sqrt{2}\right) = 4\left(\sqrt{3} - \frac{2\sqrt{2}}{3}\right)$$

Choice (4) is the answer.

10.63. In addition, from list of integral of functions, we know that:

$$\int u^n du = \frac{1}{n+1}u^{n+1} + c$$

From trigonometry, we know that:

$$\cot(x) = \frac{\cos(x)}{\sin(x)}$$

The problem can be solved as follows:

$$\int \cot(x)\sqrt{\sin(x)}dx = \int \frac{\cos(x)}{\sin(x)}(\sin(x))^{\frac{1}{2}}\,dx = \int (\sin(x))^{\frac{-1}{2}}\cos(x)dx \tag{1}$$

Now, we should change the variable of the integral as follows:

$$\sin(x) \triangleq u \Rightarrow \cos(x)dx = du \tag{2}$$

Solving (1) and (2):

$$\int u^{\frac{-1}{2}} du = 2u^{\frac{1}{2}} + c = 2\sqrt{\sin(x)} + c$$

Choice (2) is the answer.

10.64. From trigonometry, we know that:

$$\sec(x) = \frac{1}{\cos(x)}$$

$$\tan(x) = \frac{\sin(x)}{\cos(x)}$$

The problem can be solved as follows:

$$\int_0^{\frac{\pi}{3}} \sec(x) \tan(x) dx = \int_0^{\frac{\pi}{3}} \frac{1}{\cos(x)} \frac{\sin(x)}{\cos(x)} dx = \int_0^{\frac{\pi}{3}} \frac{\sin(x)}{\cos^2(x)} dx \qquad (1)$$

Now, we should change the variable of the integral as follows:

$$\cos(x) \triangleq u \Rightarrow -\sin(x)dx = du \qquad (2)$$

Solving (1) and (2):

$$-\int_{u_1}^{u_2} \frac{1}{u^2} du = \frac{1}{u} \bigg|_{u_1}^{u_2} = \frac{1}{\cos(x)} \bigg|_0^{\frac{\pi}{3}} = \frac{1}{\frac{1}{2}} - \frac{1}{1} = 1$$

Choice (1) is the answer.

10.65. From trigonometry, we know that:

$$\csc(x) = \frac{1}{\sin(x)}$$

$$\cot(x) = \frac{\cos(x)}{\sin(x)}$$

The problem can be solved as follows:

$$\int_{\frac{\pi}{6}}^{\frac{\pi}{4}} \csc(x) \cot(x) dx = \int_{\frac{\pi}{6}}^{\frac{\pi}{4}} \frac{1}{\sin(x)} \frac{\cos(x)}{\sin(x)} dx = \int_{\frac{\pi}{3}}^{\frac{\pi}{4}} \frac{\cos(x)}{\sin^2(x)} dx \qquad (1)$$

Now, we should change the variable of the integral as follows:

$$\sin(x) \triangleq u \Rightarrow \cos(x)dx = du \qquad (2)$$

Solving (1) and (2):

$$\int_{u_1}^{u_2} \frac{1}{u^2} du = -\frac{1}{u} \bigg|_{u_1}^{u_2} = -\frac{1}{\sin(x)} \bigg|_{\frac{\pi}{6}}^{\frac{\pi}{4}} = \left(\frac{1}{\frac{\sqrt{2}}{2}} - \frac{1}{\frac{1}{2}}\right) = -\left(\sqrt{2} - 2\right) = 2 - \sqrt{2}$$

Choice (2) is the answer.

10.66. From trigonometry, we know that:

$$1 + \cot^2(x) = \frac{1}{\sin^2(x)}$$

$$1 + \tan^2(x) = \frac{1}{\cos^2(x)}$$

$$\tan(x)\cot(x) = 1$$

Moreover, from list of integral of functions, we know that:

$$\int \left(1 + \tan^2(x)\right) dx = \tan(x) + c$$

$$\int \left(1 + \cot^2(x)\right) dx = -\cot(x) + c$$

The problem can be solved as follows:

$$\int_{\frac{\pi}{6}}^{\frac{\pi}{4}} \frac{1}{\sin^2(x)\cos^2(x)} dx = \int_{\frac{\pi}{6}}^{\frac{\pi}{4}} \left(1 + \cot^2(x)\right)\left(1 + \tan^2(x)\right) dx$$

$$= \int_{\frac{\pi}{6}}^{\frac{\pi}{4}} \left(1 + \tan^2(x) + \cot^2(x) + \cot^2(x)\tan^2(x)\right) dx = \int_{\frac{\pi}{6}}^{\frac{\pi}{4}} \left(1 + \tan^2(x) + \cot^2(x) + 1\right) dx$$

$$\int_{\frac{\pi}{6}}^{\frac{\pi}{4}} \left(1 + \tan^2(x)\right) dx + \int_{\frac{\pi}{6}}^{\frac{\pi}{4}} \left(1 + \cot^2(x)\right) dx = \left(\tan(x) - \cot(x)\right)\Big|_{\frac{\pi}{6}}^{\frac{\pi}{4}}$$

$$= (1 - 1) - \left(\frac{\sqrt{3}}{3} - \sqrt{3}\right) = \frac{2\sqrt{3}}{3}$$

Choice (2) is the answer.

10.67. From list of integral of functions, we know that:

$$\int x^n dx = \frac{1}{n+1} x^{n+1} + c$$

The problem can be solved as follows:

$$\int (\tan(x) - \cot(x))(\tan(x) + \cot(x))^5 dx$$

$$= \int (\tan(x) - \cot(x))(\tan(x) + \cot(x))(\tan(x) + \cot(x))^4 dx$$

$$= \int \left(\tan^2(x) - \cot^2(x)\right)(\tan(x) + \cot(x))^4 dx$$

$$= \int \left(1 + \tan^2(x) - \left(1 + \cot^2(x)\right)\right)(\tan(x) + \cot(x))^4 dx \qquad (1)$$

Now, we need to change the variable of the integral as follows:

$$\tan(x) + \cot(x) \triangleq u \xrightarrow{\frac{d}{dx}} \left(1 + \tan^2(x) - \left(1 + \cot^2(x)\right)\right)dx = du \qquad (2)$$

Solving (1) and (2):

$$\int u^4 du = \frac{u^5}{5} + c = \frac{1}{5}(\tan(x) + \cot(x))^5 + c$$

Choice (2) is the answer.

10.68. From trigonometry, we know that:

$$1 + \cos(2x) = 2\cos^2(x) \qquad (1)$$

$$\sin^2(x) + \cos^2(x) = 1 \qquad (2)$$

The problem can be solved as follows:

$$I = \int \sin(x)\cos(x)dx \qquad (3)$$

Now, we should change the variable of the integral as follows:

$$\sin(x) \triangleq u \Rightarrow \cos(x)dx = du \qquad (4)$$

Solving (3) and (4):

$$I = \int u du = \frac{1}{2}u^2 + c = \frac{1}{2}\sin^2(x) + c \qquad (5)$$

Solving (2) and (5):

$$I = \frac{1}{2}\left(1 - \cos^2(x)\right) + c = -\frac{1}{2}\cos^2(x) + \frac{1}{2} + c = -\frac{1}{2}\cos^2(x) + c' \qquad (6)$$

Solving (1) and (6):

$$I = -\frac{1}{2}\left(\frac{1}{2} + \frac{1}{2}\cos(2x)\right) + c' = -\frac{1}{4}\cos(2x) + c' - \frac{1}{4} = -\frac{1}{4}\cos(2x) + c'' \qquad (7)$$

From (5), (6), and (7), choice (2) is the answer.

10.69. From list of integral of different functions, we know that:

$$\int \frac{1}{u(x)} du = \ln(u(x))$$

The problem can be solved as follows:

$$I = \int_{\ln 2}^{\ln 3} \frac{1 - e^{-2x}}{1 + e^{-2x}} dx$$

$$\xrightarrow[\quad]{\times \frac{e^x}{e^x}} I = \int_{\ln 2}^{\ln 3} \frac{e^x - e^{-x}}{e^x + e^{-x}} dx$$

By defining the new variable, we have:

$$u(x) = e^x + e^{-x}$$

$$\Rightarrow du = (e^x - e^{-x}) dx$$

Therefore:

$$I = \int_{u_1}^{u_2} \frac{1}{u(x)} du = [\ln(u(x))]_{u_1}^{u_2}$$

$$\Rightarrow I = [\ln(e^x + e^{-x})] \Big|_{\ln 2}^{\ln 3} = \ln(e^{\ln 3} + e^{-\ln 3}) - \ln(e^{\ln 2} + e^{-\ln 2})$$

$$\Rightarrow I = \ln(e^{\ln 3} + e^{\ln 3^{-1}}) - \ln(e^{\ln 2} + e^{\ln 2^{-1}})$$

$$I = \ln(3 + 3^{-1}) - \ln(2 + 2^{-1})$$

$$\Rightarrow I = \ln\left(3 + \frac{1}{3}\right) - \ln\left(2 + \frac{1}{2}\right) = \ln \frac{10}{3} - \ln \frac{5}{2} = \ln \frac{20}{15} = \ln \frac{4}{3} = \ln 4 - \ln 3$$

$$\Rightarrow I = 2\ln 2 - \ln 3$$

Choice (3) is the answer.

In this problem, the rules below were applied.

$$e^{\ln a} = a$$

$$e^{-\ln a} = e^{\ln a^{-1}} = a^{-1}$$

$$\ln a - \ln b = \ln \frac{a}{b}$$

$$\ln a^n = n \ln a$$

References

1. Rahmani-Andebili, M. (2021). Calculus – Practice Problems, Methods, and Solutions, Springer Nature, 2021.
2. Rahmani-Andebili, M. (2021). Precalculus – Practice Problems, Methods, and Solutions, Springer Nature, 2021.

Index